QUANTUM
CHRISTIAN
REALISM

QUANTUM

CHRISTIAN

REALISM

HOW QUANTUM MECHANICS UNDERWRITES AND REALIZES CLASSICAL CHRISTIAN THEISM

ROCCO BONI

WIPF & STOCK · Eugene, Oregon

Wipf and Stock Publishers
199 W 8th Ave, Suite 3
Eugene, OR 97401

Quantum Christian Realism
How Quantum Mechanics Underwrites and Realizes Classical Christian Theism
By Boni, Rocco
Copyright©2016 by Boni, Rocco
ISBN 13: 978-1-5326-8606-1
Publication date 3/22/2019
Previously published by XLibris, 2016

Unless otherwise indicated, all scripture quotations are from The Holy Bible, English Standard Version® (ESV®). Copyright ©2001 by Crossway Bibles, a division of Good News Publishers.

Used by permission. All rights reserved.

CONTENTS

Preface xiii
Introduction xix

Chapter 1 Quantum Mechanics and Christianity 1
Chapter 2 The Province of Providence 28
Chapter 3 Interlude: The Intelligible Cosmos 55
Chapter 4 The Medium of Divine Mind 60
Chapter 5 The Mechanism of Divine Action 96
Chapter 6 Quantum Uncertainty and the Subtlety of God 117
Chapter 7 How to Design a Christian Universe 130

Index 139
Bibliography 147

To my blessed and beloved mother Gloria

Her children rise up and call her blessed; her husband also, and he praises her: "Many women have done excellently, but you surpass them all."

Proverbs 31:28-29

For almost three centuries the Christian faithful have waited in uncomfortable silence for science to report something tangible as to the existence of God.

Quantum physics has now broken that silence.

This work is an account of what it has said.

Tell me, since you have no doubt that the world is ruled by God, do you know *how* it is governed?

—Lady Philosophy to Boethius

Preface

Is God an Active God?

I have always been interested—or more accurately perplexed—as to how God acts in the world. It has long seemed obvious to me that since God's actions are not *overt*, that they must be *covert*. Of course, in a physical universe there are only so many ways for an agent to act upon a material object so as to physically effect and/or influence it. Generally speaking, you can only either push or pull a material object; whether it be microscopic or macroscopic—whether an atom or a billiard ball. Whatever ones purposes of action may be, ones kinetic options are rather limited.

Human agents then, can exert a physical force upon matter by either pushing or pulling. But is God, as an active agent in nature and the world, limited by the same? Is this how God physically moves matter about? Literally by pushing and pulling? Is God like an omnipresent billiards player knocking about the constituent particles of the universe? And if so, (pressing the analogy) what does He use for a cue stick? That is, in what way does God impart the push/pull force? How exactly does God come into physical contact with matter so as to physically influence it? By what means does God set matter in motion?

Such questions were at the forefront of the Enlightenment model of divine action; which, model, was "interventionistic." Perhaps not surprising, considering the model arose against the backdrop of Newtonian physics, which imagined the cosmos as a rigidly law-governed (and hence causally-closed) system that is both deterministic and reductionistic. Divine action as 'causal intervention' was thus inescapable under the Newtonian ontology.[1]

---

[1] Newtonian mechanics gave birth to deism which re-imagined God as a disinterested (perhaps even absentee) superintendent of natural law and process with no interest in providential guidance of any kind. Contra the God of theism, which is both personal and active, the god of deism has interest in neither mankind nor the world, and hence cause for neither miraculous intervention nor particular providential action. Newton (1642-1727) however, despite having discovered the three laws of determinism (which subsequently led to the Enlightenment ideal of a causally-closed world ontology), remained an avowed theist. Newton for instance asserted that due to the relentless tug of gravity between the planets God would necessarily have to intervene from time to time to correct for orbital precess—such as in the perihelion of mercury. Not all however were impressed with Newton's *ad hoc,* god-of-gaps-style solution. Gottried Wilhelm Liebniz (1646-1716) for instance—no friend of Newton's—criticized Newton for portraying God as a 'divine tinkerer' too incompetent to get things right at the first. And in 1787 Pierre-Simon Laplace (1749-1827)—no friend of divine intervention—claimed to have mathematically resolved the celestial instability dilemma—and with it the need for a God hypothesis (or perhaps more accurately, Newton's "god-of-the-gaps" hypothesis). Laplace's mathematical resolution however later proved to be incorrect. Ultimately, it would be Einstein's Theory of General Relativity that resolved the Newtonian dilemma. This historical anecdote is important for a number of reasons, not least because it marks the beginning of the modern debate over the nature of divine action (see for instance the Leibniz/Clarke correspondence which began in 1715 and continued until the death of Leibniz in 1716. (See, for instance, H.G. Alexander, *The Leibniz-Clarke Correspondence: Together with Extracts from Newton's Principia and Opticks,* [Manchester University Press; Manchester, 1998]). Up until this point in history debates in providence were predominantly theological, rather than scientific; ranging from issues of God's benevolence (Marcus Minucius Felix fl. c. 150-270 AD., Lactantius 240-320 AD.), and God's justice (Irenaeus 130-202 AD.), to issues of God's judgment and Grace (Augustine 354-430 AD.), to issues of God's sovereignty *vis-a-vis* human free will (Chrysostom 347-407 AD., Pelagius 390-418 AD., Thomas 1225-1274 AD., Melanchthon 1497-1560 AD., Calvin 1509-1564 AD., Molina 1535-1600 AD., Arminius 1560-1609 AD., et al.). For the first sixteen centuries of Church history, talk of divine action tended toward talk of God's general providential governance of nature and nature's laws—and whether or not this governance extended to human action. Divine Sovereignty and the extent thereof tended to be the major issue at hand. Issues relating to theological determinism, predestination, foreknowledge, foreordainment, compatibilism, incompatibilism, omnicausality, unilaterality and the like, was common grist for theological debate. Such topics are of course still popular today but a shift of sorts has taken place, a shift away from general providence and toward particular providence (also called special providence). Since the scientific revolution and the birth of Newtonian mechanics in particular (not to mention the Cartesian paradox) concern has turned to the question of the *causal* nature of divine action and/or the *mechanics* of the God/world interface. However, it was soon realized that the questions raised by Newtonianism could not be answered by Newtonianism (that is, not without scientific and theological paradox). Subsequently, by the 19th century, special providence had fallen by the wayside. Along with the notion of a personal God. Science (and classical mechanics in particular) simply could no longer abide the interventionist God of Judeo-Christianity. Theism, subsequently gave way to deism; and deism to atheism.

Today, things have changed radically. Since the commencement of the twentieth century, physics has advanced far beyond the Newtonian worldview—not only 'relativistically', but *quantum mechanically!* With the latter ushering in a radical new ontology, thoroughly antithetical to the Newtonian one.

The early twentieth century witnessed the birth of the quantum revolution which radically altered our understanding of not only cause-and-effect but of physical matter itself! In consequence, the Newtonian ontology was pronounced dead, along with the crude image of God as an invisible particle-pusher physically knocking atoms around.

Today, in place of the old reductionistic, deterministic Newtonian ontology we have the new quantum ontology. And with it a new model of divine action. No longer does the disembodied God of theism have to somehow temporarily don physicality(?) in order to physically influence the material constituents of the world. For, according to recent insights reality is far less physical then we had previously imagined. Moreover, this *absence of physicality* in nature appears at precisely the right level—and in precisely the right manner—and to precisely the right extent, for to allow the disembodied Deity of Judeo-Christianity to *causally* engage, and thus *effectually* influence the physical world—albeit *incorporeally!*

Reality thus appears to be tailored towards the reality of a disembodied, active Deity! This is wonderful news for theists who have long believed in God's active involvement in nature, history, and the world. Regarding then, the above question; "is God an active God?", the answer appears to be a resounding yes!

However, establishing that God acts, only begs the question: "why doesn't God act more often?" This is the question of questions, such that any claim as to the reality of divine action must at some point humbly address the reality of God's seeming absence of action.

Where is God?

For millennia men have brooded over the extent of God's presence in nature, history, and the world. And for millennia men have languished over God's seeming absence. Often wearily, believers have kept the faith, as instructed in Scripture. Through valleys of shadow, darkness and death we pray to a Deity we cannot always feel is there. Indeed, who we may rarely feel is there. This is the human condition. And it is one of faith. And through faith, hope.

We may not understand why many things happen—or more honestly, *why God allows many things to happen*. And we may not understand why *this world* rather than some other less horrific world—the universe was after all a *contingent* creation, and so presumably God could have created any kind of world He so desired (key word here being "presumably"[2]).

The frustrating fact of the matter is that we simply do not know why God did things the way He did (or does things the way He does). And so at the end of the day, all of our questioning, judging and second guessing of God must ultimately yield to trust. Trust that God has very good reasons for doing whatever He does, even if it appears senseless to us. It seems we must not only trust God, but trust that we can trust God.

Regarding the problem of evil and the question of why God allows bad things to happen to good people, I suspect that we today are in a position not unlike the apostles as they watched their messiah being nailed to a Roman cross. This would have been as hurtful as it was perplexing. For, at the time, nothing could have seemed more contrary to the perceived will of God than the long awaited messiah's public execution.

Those who lose faith and trust in God in the face of the world's ills, evils and trials are not unlike John the Baptist, who, sitting despondent in Herod's dungeon, sent his disciples to ask Jesus "are you the one who is to come, or should we look for another?" The reply, in essence, was that Jesus was indeed the long awaited messiah. John and the disciples therefore continued in their expectant hope for the long awaited re-establishment of the Davidic throne. Their 'expectant hope' however was in vain. For, God, as seems always to be the case, had other plans. I ask then; was John's beheading unexpected? Absolutely. Was Jesus's Crucifixion incomprehensible? Most assuredly. Yet God was ahead of the problem, working all things for the greater eternal good of all? (Which I suspect, is always the case).

---

[2] I am not here questioning God's omnipotence, but rather, whether or not the creative end goals desired by God could have been achieved without the high cost of human and animal suffering. Perhaps the creative end goals desired by God necessitates just this sort of universe. The point is that we simply do not know certain things—at least not yet!

If there is one thing the New Testament gives us, it is hope; along with incentive to trust God. Incentive to trust that God has an answer too and for the worlds evils—a response to the fallen nature of the world. We must *and can* trust that God is again, ahead of it. That God will one day make things right. If there is one thing that Scripture testifies unto, it is that God *always keeps His promises*. And God's promise is to one day "wipe away all tears." We can therefore be confident in our trust, and take comfort in our hope.

And so while there are many things about the nature of this world that we cannot understand, we can at least trust that a divine response and/or reckoning awaits. In the meantime, the believer can know that God is indeed there, able and active; despite the ambiguity of the latter category. *Which*, category brings us to one of the central topics of this work: divine action. Believers have long struggled to fathom the God/world relationship; to believe that God is truly there with us in our joys, sorrows, and griefs.

Until very recently in history[3] Christians had only their faith to reassure them of God's presence and action in the world. Then, early into the twentieth century the quantum revolution happened, and nature chimed in on the subject with a testimony so incredible, and so theologically charged, as to be in seeming miraculous cahoots with the Scriptural testament of a "theistic" God—*that is*, a present, personal, *active* God!

With quantum physics came a revolution in our understanding of the very nature of physical reality itself. Quantum mechanics demands a reconceptualization of physical reality. One wherein physical reality *isn't physical at all!* Gone forever are the days of classical material monism. Quantum physics has revealed that reality is dualistic, being only 'physical' in part—namely, at the macro-level of reality.

The cosmos, as we now know it, is mysteriously physical and/or material at the uppermost level of the ontological hierarchy of reality while being mysteriously non-physical and/or immaterial at the bottommost level of the ontological hierarchy of reality. This is quantum dualism and it divides reality into that which is physical, and that which is non-physical. Understand that this "non-physical" aspect of reality isn't the same as "nothing", rather, it is a non-physical "something"; an

---

[3] Post quantum revolution (i.e., mid-twentieth century to present).

immaterial substance of some sort. It even has an intelligible structure; a dynamic structure, and can be mathematically modeled quite accurately. Oddly enough, this structure is wave-like. But remember this wave-like structure is non-physical. It is ghost-like. Or, as I suspect, 'spirit-like.'

Whatever this substrate, it is uniquely, indeed ideally suited for interfaction with a platonic-like essence or mind.[4] What this means is that the physical universe is engineered so that a transcendent "mind" can interface with its physical constituents in a purely "cogitative" way! Gone is the inelegant particle-pusher model which imagines God as a celestial Billiardsman brutishly knocking particles about.

What reality has revealed is a system far more elegant and far more economical. By and through the laws of quantum mechanics, God (whom recall is a disembodied Spirit [John 4:24]) can simply "will" matter about. But it gets even better than that, insofar as God can do all this without infringing upon natural law and/or the governing laws of science! For, it all takes place within the parameters of what quantum law will allow!

For the first time in history we have actual scientific insight into the reality and nature of the God/world relation! So while we may not yet be able to fathom why this or that happens, we can be *confident* that God is indeed there, lurking just behind the physical scenes in power and glory prepared to make things right. In this "confidence" we can root our trust. And in this trust, our hope.

---

[4] Platonic triangles may be eternalistic, but abstract wave-functions are temporal and dynamic; not unlike minds, thoughts, and consciousness. Wave-functions are also stochastic (lending to their ontic openness) and thus have the capacity to be altered from either within or without. Such as, by the separate probabilities of overlapping wave-functions, or else by the basic "determining" actions of the divine Mind itself, which, after all, is a disembodied consciousness.

Introduction

The Problem

According to mainstream science we live in a physical universe that is mechanistic and materially monistic. According to Newtonian physics (and the deterministic ideal thereof), all action is governed by lock-step cause-and-effect processes that span backward in time all the way to the initial prime movement (or initial cause) that set everything in motion. These cause-and-effect processes are in turn governed by universal, immutable laws—scientific laws. And from these laws derive the faithful regularities of nature.

Scientific materialists see the universe as a closed cosmic system wherein all life is ultimately doomed to entropic extinction. According to these, the physical universe is the whole of reality; there is nothing more, nothing less. In the notoriously bleak words of Carl Sagan, "the cosmos is all there is, all there ever was, and all there ever will be."[5] Such is reality to the scientific materialist. Within such a reality mankind holds no place of significance. We are simply part of the vast cosmic machine, born of its stardust, bound by its laws, fated to its finitude—mere biological cogs, caught-up within the causal wheelworks of the great cosmic clockwork.

To the materialist, all that occurs, occurs naturalistically—that is, *mechanistically*. This holds true for both the inorganic realm as well as

---

[5] Carl Sagan, *Cosmos*, (New York; Random House, 1980).

the organic realm—including the biological realm; including us. What we experience, we experience physically—that is, *sensorially*. This means chemically, and hence mechanistically.

René Descartes (1596-1650) long ago reasoned that beasts are material monist machines made out of meat. Humans however were seen as something far more interesting. According to Descartes man is a dualistic being possessing both a material body and an immaterial soul. Demonstrating the latter, however, proved futile. Descartes, along with a good deal of Enlightenment thinkers, searched in vain for the locus of the elusive "soul substance." But the Cartesian soul substance was never found and so Enlightenment science came to conclude that man too is but a material monist machine—a biological clockwork.

By the 19th century, the clockwork universe model was universally accepted scientific dogma. Free will[6] was deemed an illusion and all life was judged to be mere mater in motion. In one fell swoop, all that is man was reduced to random chemical reactions taking place within the gelatinous aggregate of grey-matter known as the brain. Our most noble actions, our most profound insights, our most sublime experiences—all that we are, all that we hope to become, all that we feel—none of it means anything. All is an illusion, a mental mirage, being no more meaningful than any other chemical reaction in nature. All there is: atoms and the void. This is the conclusion of the modern materialist scientific paradigm, as deduced from Enlightenment thought.

Such a conclusion, however, begs a number of metaphysical questions. For, if we truly live in a material monist reality, what then, of God? What then, of spirit? What then, of soul? Illusions we are told; evolutionary vestiges from mankind's cultural childhood. Firmly refuted remnants from a time when we invoked the gods to give meaning to nature's caprice and to better cope with the vagaries of life.

But today, we are told, we know better. Today, we know that even our most sophisticated superstitions are still 'just superstitions.' Today, we know that belief in the divine, as comforting as it may be, is simply not the reality. The reality is "physicality" plain and simple. There is no "spirit." There is no "soul." There is no "God." There is only atoms and the void. Matter, energy, and force phenomena: this is the real trinity,

---

[6] I refer here to *libertarian* free will (i.e., genuine free will, which affirms both volitional freedom and agent-causation), and not *compatibilist* varieties which account for human actions deterministically.

and it is all there is, all there ever was, and all there ever will be. It is time for mankind to grow up and face the inevitable—i.e., the obvious: we are alone; mere flukes of matter in a brute-fact universe.

This is the story repeatedly told to us (*and continually sold to us*) by the contemporary scientific paradigm, which, being rooted in Enlightenment ideology is materially monistic, and hence intrinsically atheistic.

The problem however, with the atheistic conclusions of modern science is that it presumes the deterministic ontology of Enlightenment physics, that is, Newtonian physics. Newtonian physics is today referred to as "classical physics" for two very good reasons: the first being *relativity*, the second being *quantum mechanics*, both of which were developed and set forth in the first half of the twentieth century. You see, the atheistic conclusions of contemporary science are founded upon the now-discredited "deterministic ontology" of classical physics, when in reality, present-day physics is a uniquely indeterministic, and hence *post-enlightenment* paradigm!

Quantum physics radically undercuts the material monist metaphysic that has been, *and is presently being* smuggled into the *so-called* "objective" scientific worldview. The reality is that quantum physics, with its *indeterministic* ontology has totally revolutionized our picture of the world. No longer do we live in a causally closed 'material' world system where there is place for neither God, nor His providence, nor for soul or spirit, nor for volition, or free will. No longer do we live in a mechanistic, monist reality wherein all that exists is the physical, the natural, the material. Today we live in a new reality—*a quantum reality*. And contrary to the Newtonian/Enlightenment schema, it is spiritually robust. So welcome to the new reality; it is uncannily theistic—*and particularly Christian.*

The Solution

This work is an account of the fascinating implications of contemporary quantum physics for classical Christian theism—with particular attention paid to the implications of quantum indeterminacy for divine action. More generally, this work sets out to demonstrate how quantum mechanics underwrites the central tenets of classical Christianity while resolving a number of longstanding disparities between science and Christianity—as well as a number of longstanding theological disparities internal to the faith.

On the whole, this work might be said to be a physics-based account of the '*mechanics*' of Christian theism. Or more specifically, the "quantum" *mechanics* of Christian theism. For, as we shall see, quantum mechanical process and the phenomena thereof strongly suggest that we live in a universe that has been ontologically engineered to underwrite, actuate, and ultimately realize the theologically-based tenets of classical Christianity. Here then, is your own personal guide to the (uniquely unseen) 'physics of the faith.'

It is my hope that by demonstrating that the key tenets of Christianity are grounded in the quasi-physical processes of quantum mechanics, that believers will be edified in both their faith, and the objective reality thereof, unintimidated and undeterred by the pervasive materialism that so dominates our present culture.

Chapter One

Quantum Mechanics and Christianity

> If quantum mechanics hasn't profoundly shocked you, you haven't understood it yet.
>
> —Niels Bohr

Introduction

Science occasionally makes a discovery so profound and so unprecedented that it radically challenges and ultimately changes our conception of reality. Some discoveries are in fact so profound and so unprecedented that their ultimate implications escape notice early on, requiring a span of time for the human mind to acclimatize unto its truths.

Just such a discovery has been made; and this work is the telling of its truths. The discovery itself is rooted in the phenomena of quantum mechanics and its implications are revelational to say the least. As it turns out, the central tenets of quantum physics pander to the central tenets of the Christian faith!

Never before has science spoken so clearly or so plainly on the reality of the divine, or provided so detailed an account of the mechanics of spiritual activity within the physical world. Quantum physics has provided us with an unambiguous look at the *ultimate reality* behind what we experience as the *physical reality* of mere appearances. And quite amazingly, this 'ultimate reality' turns out to be uniquely and unapologetically "Christian" in its ontological standing. This work is the account *of that standing*; and what it reveals the ultimate nature of reality to be.

Let us then begin our journey into this "ultimate reality" with a preliminary consideration of the physics that takes us there: *quantum physics*. Quantum physics, or 'quantum mechanics' as it is formally referred to, is an incredibly heady subject that very few laypeople will have encountered in any substantive way. A requisite (non-technical) overview of key quantum phenomena is therefore required for those readers unfamiliar with the theory's intrinsic eccentricities.

Let us then begin at the beginning—which, as any physicist, philosopher, or theologian knows, is *with a question*. Or in this case, two questions:

- *What exactly is quantum mechanics?*
- *And how can it possibly relate to God and Christianity?*

Brief

Simply put, quantum physics is the study of the nature of reality at its smallest scales. More specifically, it is the study of the interactions of matter and radiation at the subatomic level. Among other things, quantum physics has revealed that energy is composed of discrete, indivisible units dubbed "quanta." Quantum physics thus concerns itself with the nature and mechanics of these smallest units of indivisible energy. Energy is thus said to be "quantized."

Quanta and quantum particles are the fundamental building blocks of the physical universe, and mysteriously, indeed paradoxically, they can exist as either "waves" or "particles." The waves of quantum-level reality however are not ordinary 'physical' waves such as we are accustomed to at the ordinary everyday level of macro-reality. Rather, the waves of quantum-level-reality are *non-physical,* being both *intangible* and *incorporeal*. It can thus be said that all physical reality is reducible to incorporeal quantum waves, and/or quantum-wave-systems known as "wave-functions."[7] And as we shall see, these mysterious quantum waves

---

[7] Formally, a quantum 'wave-function' is a mathematical (i.e., probabilistic) registry of a particles quantum state with respect to position/momentum/time/and/or spin. Informally, the term 'wave-function' is used to refer to the (objectively existent) abstract wave-like reality that is the solution to the Schrödinger wave-equation.

are the key to fathoming the inscrutable relation between spiritual reality and physical reality.

As it turns out, the spiritual reality that the faithful have long felt in their hearts and believed in their minds necessarily overlaps physical reality, and quantum physics provides us with a never before dreamt of glimpse into the nature of its workings. But before we get into the 'mechanics' of spiritual reality (and Christian theism[8] in particular) we must better acquaint ourselves with the curious nature of *quantum reality*—which, as already mentioned, is the key to relating the physical and the spiritual, and resolving the host of conundrous scientific and theological issues that have so long beset classical Christianity.

From a secular scientific perspective quantum physics is simply the study of the laws and principles that govern quantum-level process and phenomena. In this work, our primary focus *will not be* on the mathematical formalisms of these laws and principles, but rather, on the nature and implication of the processes and phenomena that these quantum laws and principles give rise unto.

For our purposes, it is more important to grasp the *concepts* of quantum theory than the calculations of quantum theory. For, it is here, within the 'concepts' of quantum theory (and within the anonymity of quantum uncertainty in particular) that naked reality is most clearly seen, and the foundational ontology[9] upon which the whole of physical reality rests is most clearly revealed.

At the heart of the quantum dilemma is the question: what is the metaphysical/ontological status/nature of the quantum-level entity we dub "wave-function." One thing we know is that whatever wave-functions are, they are abstract, mathematical, indeed nomological, with an incorporeal wave-like structure that is objectively existent and causally efficacious.

[8] The term "theism" comes from the Greek *theos* meaning 'God' and is used to refer to a very particular deity in the western world; namely, the God of western Monotheism (i.e., the Personal, Creator God of the Abrahamic faiths). This God is believed to be intimately concerned with humanity and the world (particularly in Judeo-Christianity where God dies to redeem both man and world; both creature and creation, back unto himself). It is from the word *theos* that we derive terms like *theology, theodicy* etc., The close association between the ancient Greek *theos* and the western Judeo-Christian Deity traces back to the terms first historical appearance in Plato's *Politeia* (*Republic* II, 397a), and subsequently in the writings of Aristotle (*Metaphysics* III, 4, 1000 a 9), from whence it made its way into medieval Christian thought *via* Thomism (Jammer, *Einstein and Religion: Physics and Theology* [Princeton New Jersey: Princeton University Press, 1999]).

[9] *Ontology* is the study of the nature of "being", "existence", and/or "reality", and concerns the question of how reality actually is, in itself, objectively, independent of human conception. The ontological question of 'how things actually are', is therefore quite separate from

Matter of Mind

Incredibly, to the dismay of classical physicists (and chagrin of material monists), the ontological foundations of quantum reality, which serves as the very ground of all things physical, turns out to be indistinguishable in essence from "rational abstraction." As it happens, the fabric of physical reality is *not physical at all*. In fact, that which serves fundamentally, foundationally, and elementally as the basis of physical reality is more akin to 'abstract cogitation' than anything else.

However, unlike genuine cognition (which is a *subjectively* existent phenomenon), the quantum essence is an *objectively* existent phenomenon. This is indeed a "hard saying." Nonetheless it is a fact of contemporary physics. Put plainly, quantum wave-functions are objectively existent abstract mathematical systems! However, unlike other equation-based mathematical systems (which may or may not be considered within the subjective thoughts of any particular mathematician) wave-functions, while similarly accessible to the minds of mathematicians abstractly and mathematically via the Schrödinger wave-equation, nonetheless also exist independently at or within quantum-level reality! Having said this, let me say that I am not here speaking simply 'platonistically.'[10] You see, wave-functions are indeed platonic entities, but they are also *trans*-platonic entities, in that they exist both within platonic reality and *beyond* platonic reality—specifically, at or within the *quantum foundations of this physical universe!*

It is baffling to the mind to attempt to fathom how this can possibly be so. Yet the fact remains: it is so. Wave-functions then, are not simply 'mathematical ideations', they are *objectively existent* 'mathematical ideations.' Moreover, these mathematical ideations, while purely abstract, nonetheless possess a very particular nomological structure being both formally and dynamically wave-like.

---

the epistemological question of 'how things merely appear to us', either as subjectively perceived, or as altered via our methods of observation.

[10] Philosophers today distinguish between "Platonism" (with a capital "P") and "platonism" (with a lower case "p"). "Platonism" (capital P) is used to refer to the entire philosophical system of Plato, whereas "platonism" (lower case p) is used to refer merely to the belief in the existence of abstract objects. Throughout this work I advance a form of mathematical platonism (lower case p) wherein the mathematical forms themselves directly realize quantum process.

Quantum physics reveals that physical reality literally reduces to a mentalistic essence not unlike that of 'thought', or 'conception.' Beneath all physicality there exists a foundational quantum substrate that is wholly incorporeal and akin to the conceptual slate of cognition whereupon the rational mind does its bidding. As Sir Arthur Eddington simply but straightforwardly put it, "The stuff of the universe is mind stuff."[11]

The discovery of quantum mechanics during the early to mid-decades of the twentieth century shocked the classical physicists of the day by revolutionizing our conception of reality.
Quantum mechanics has literally put us into contact with a metaphysical reality not unlike the abstract realm of eternal forms first fathomed by Plato in the fourth century BC.[12] Can quantum physics, and more specifically the *'quantum wave-function'* be the first rays of light beyond the cave? I believe so, and the implications for our understanding of classical Christian theism are profound, if not revolutionary. Equally profound and revolutionary are the destructive implications of quantum mechanics for atheism and its primary assumptions of material monism, determinism and reductionism.

Quantum physics, insofar as it reveals a dualist ontology (wherein classical reality is distinctly physical and quantum reality is distinctly *meta*physical) strikes at the very heart of secular scientistic atheism by quashing its core assumption of material monism. What's more, quantum theory at once defeats metaphysical reductionism (and its atheistic implications) by demonstrating that physical reality ultimately reduces to, and arises from, an ethereal bedrock of incorporeality, the closest analogue of which, is consciousness. Scientific materialism simply cannot stand against the quantum fact that material reality ultimately emerges from immateriality. Quantum pioneer Niels Bohr, perhaps put it best when he stated that "Everything we call real is made of things that cannot be regarded as real."[13] Here again we are reminded

---

[11] Sir Arthur Eddington, *The Nature of the Physical World* (Cambridge University Press; Cambridge, 1948).
[12] See Plato's *Allegory of the Cave* in Book VII of the *Republic* (circa 380 BC), and also Plato's *Theory of Forms* in the *Timaeus* 49cff. (Circa 360 BC).
[13] Bohr, to Einstein, at Fifth Solvay Conference on Quantum Mechanics, 1927.

of platonic thought.[14] Like the Pythagorean triangle of the mind's eye, the quantum wave-function is abstract, noumenal and transcendent. Yet unlike the Pythagorean triangle it is also immanent (and objectively so!)

The quantum substance, whatever it is, is unmistakably platonic, being strikingly akin to Plotinus' *nous,* or *soul essence* as found within the neo-Platonic hierarchy of Being.[15] Understand that the platonic (i.e., abstract/ideational) nature ascribed unto quantum wave-function is not simply metaphysical fluff; rather, it is hardcore, empirical, *indeed demonstrable,* scientific fact.[16] What's more, it is "orthodox" scientific fact.[17] There is an astonishing degree of correspondence between quantum reality and platonic reality. So much so, that Heisenberg, who had read Plato's *Timaeus* as a teenager, openly acknowledged the ontic sameness of the two realms.

> I think that modern physics has definitely decided in favor of Plato. In fact the smallest units of matter are not physical objects in the ordinary sense; they are forms, ideas which

---

[14] Particularly "mathematical platonism", which says that mathematical realities are objectively existent abstract realities that are transcendent, eternal, immutable. Influential modern and contemporary platonists include, Gottlob Frege, Kurt Gödel, Bertrand Russell, W.V.O. Quine, Alain Connes, René Thom, Roger Penrose, John Leslie, and Max Tegmark, to name a few. What's more I should point out that while throughout this work I refer to quantum reality and wave-function as being "platonistic" and/or "platonic-like", I do not mean to suggest that either quantum reality or wave-function has existed eternally *relative to some generic demiurge,* but only that quantum reality and wave-function (as created by the Judeo-Christian God) seem to be ontologically and substantively similar to the mathematical realm of forms described by Plato. The mathematical platonism spoken of throughout this work then, is of a Christianized variety, with the realm of mathematical ideals corresponding to perhaps a quasi-spiritual or divine sphere.

[15] Neo-Platonism was a system of philosophical thought developed by Plotinus in the 3rd century A. D. Neo-Platonism was largely an amalgamation of mysticism and Platonism with elements of Judaism, as well as early Christian thought.

[16] The single-particle interference patterns that arise in modern double-slit experiments clearly and powerfully demonstrate the objective reality of wave-functions.

[17] The trans-spatiotemporal nature of quantum reality is perhaps best demonstrated in the famous delayed-choice experiment of John Wheeler, wherein it is shown that quantum potentialities in the *past* can be affected by quantum events in the *present.* Put plainly, it is now possible to (in limiting cases, of course) alter the history of a physical event, *from the present*! What is more, delayed-choice/interferometer experiments prove that quantum wave-functions, though *wholly incorporeal* are nonetheless *objectively real!* This is of paramount importance to the science/God question in that the scientific discovery of an objectively existent incorpreality brings us one step closer to realizing the spiritual nature of ultimate reality.

can be expressed unambiguously only in mathematical language.[18]

Heisenberg here describes the quantum ontology as being indistinguishable from the platonic ontology—which, of course, is a *metaphysical dominion of ideals*—or, in the quantum case, a *metaphysical dominion of abstract mathematical constructs*.

Enchanting is the reality.

Creation, Christianity, and the Quantum

The depiction of wave-functions as objectively existent incorporeal mathematicisms was vehemently resisted by the classical (Newtonian) physicists of the day. According to these classicists, (most notably Albert Einstein and Erwin Schrödinger) wave-functions (which are solutions to the Schrödinger wave-equation) are but mathematical approximations to objectively existent, *physical* wave systems every bit as tangible as sound-waves and water waves. Yet, despite this commonsense conclusion of Einstein and Schrödinger, all the evidence suggested (and continues to suggest) that quantum wave-functions are objectively existent, *incorporeal* wave-systems every bit as *intangible* as dream states and platonic entities—just as Heisenberg had described.

To the shock of physicists, and the chagrin of material reductionists everywhere, the bottom-most layer of the ontological hierarchy of the physical universe isn't 'physical' at all, but is literally "quantum wave-function." This is the greatest scandal in the history of physical science in that wave-functions, being mathematically constrained mental constructs, have an objectively real existence![19] More incredibly yet, these abstract mathematical constructs literally actuate into being the physical constituents of which the material universe, *and you and I*, are ultimately made!

Wave-functions then, are absolutely fundamental to the existence of the physical universe. What's more, they are absolutely fundamental

---

[18] Werner Heisenberg, *The Development of Quantum Mechanics*, Nobel Lecture, December 11, 1933.

[19] We know this because quantum wave-functions can and do generate objective, observable phenomena (e.g., electron diffraction double-slit interference patterns, etc.).

to the implementation and execution of the core tenets of the Christian faith. Incredibly then, both the *physical* and the *spiritual* foundations of creation, owe directly to the dualistic nature of the quantum substrate!

Quantum Dualism

Quanta and/or quantum particles are dualistic entities which can manifest as either localized physical particles or as incorporeal non-localized waves. When a quantum particle (or particulate system) ceases from being observed (and/or interacting with macro-level matter aggregates [such as measuring devices]) it instantaneously assumes the incorporeal status of wave-function. Conversely, when an incorporeal wave-function is observed (and/or interacts with a macro-level matter aggregate) it instantaneously actuates as a physical quantum particle. Hence quantum dualism.

Quanta (e.g., photons, electrons, etc.,) can exist as either individual particles or as particle-systems (atoms, molecules, etc.). These, when in wave-form (*as 'wave-functions'*) exist as mathematically governed incorporeal entities, and evolve stochastically over time, generating statistical probabilities as to the likeliest places, properties, values, etc., for which to manifest upon physical re-actuation.

A quantum wave-functions then, is no less than a potentiality catalogue marking all the possible and/or potential states of being and/or existence that a quantum particle or system might assume upon its physical actualization (which, recall occurs when the wave-function once again engages and/or interacts with physical reality).[20]

---

[20] Heisenberg keenly recognized that wave-functions, as abstract mathematical systems, are unequivocally platonic. Similarly, Heisenberg did not fail to recognize that wave-functions, as stochastically governed probability densities, are uniquely reminiscent of the Aristotelian concept of "potentia." Heisenberg writes; "In the experiments about atomic events we have to do with things and facts, with phenomena that are just as real as any phenomena in daily life. But the atoms and the elementary particles themselves are not as real; they form a world of potentialities or possibilities rather than one of things or facts… The probability wave…means a tendency for something. It's a quantitative version of the old concept of *potentia* in Aristotle's philosophy. It introduces something standing in the middle between the idea of an event and the actual event, a strange kind of physical reality just in the middle between possibility and reality." (Werner Heisenberg, *Physics and Philosophy* [New York: Harper and Brothers 1958]). Even in his ontic statement on the matter Heisenberg incorporated Aristotelian thought; writing: "The probability function combines objective and subjective elements. It contains statements about possibilities or better tendencies ('potentia' in Aristotelian philosophy), and these statements are

Of course, the question of exactly *how* a quantum wave-function, which is literally an abstract entity, can possibly engage with, and/or interact with physical reality is a fundamental (and ontological) mystery of quantum mechanics unto which there is no answer (nor can there ever be one). All we know for sure is that quantum waves (which are non-physical) become quantum particles (which are physical), and quantum particles (which are physical) become quantum waves (which are non-physical).

Within this foundational truth of quantum physics we discover one of the central tenets of quantum theory: *ontological dualism*. The fundamental constituents of which physical reality is constituted (including you and I) can exist as either physical/material particles (electrons, protons, neutrons, photons[21], etc.) or else as non-physical/immaterial mathematical abstractions (i.e., wave-functions), such that the quantum realm is marked by a unique brand of either-or dualism. Ultimately then, macro *materiality* emerges from quantum *ethereality*. That is, from "quantum wave-function."

The theistic implications here, are quite clear. The physical universe is continually emanating from an elemental substrate with distinctively mind-like properties. I say 'mind-like' because the quantum wave-function, as an abstract store of existential possibilities, comprises a sphere of volitional freedom from whence choices (in the form of quantum event outcomes) derive.[22] This is highly suggestive in that abstract *possibility*, coupled with non-physical *freedom of action* and

---

completely objective, they do not depend on any observer; and it contains statements about our knowledge of the system, which of course are subjective in so far as they may be different for different observers. In ideal cases the subjective element in the probability function may be practically negligible as compared with the objective one." (ibid.).

[21] Photons have zero mass so technically they are not 'material' particles. However, they do have energy and so technically they are 'physical' particles. In this way photons are physical but not material.

[22] Quantum systems (e.g., atoms) are able to make seemingly conscious choices from among indeterministic states. Freeman Dyson states: "Atoms in the laboratory are weird stuff, behaving like active agents rather than inert substances. They make unpredictable choices between alternative possibilities according to the laws of quantum mechanics. It appears that mind, as manifested by the capacity to make choices, is to some extent inherent in every atom." (Freeman Dyson, Templeton Prize Address, "Progress in Religion", 2000). Of course, the question is whether this seeming 'consciousness of choice' belongs to the quantum level constituents themselves, or else to a divine Mind behind, acting in and through quantum level process.

incorporeal *choice*, are together the exclusive function of mind (or perhaps, in this case "Mind").

Quantum Dualism and Christianity

As we will see, quantum dualism is brilliantly suited to realize a very specific set of theological doctrines and truths—namely, those particular to Judeo-Christianity. Among these, are the doctrine of the soul, the doctrine of libertarian free will (and/or human volition), the doctrine of *creation continua*, the doctrines of general providence and special providence (the former [general providence] refers to God's initial act of creation and His subsequent act of upholding and sustaining the world in its continued existence, while the latter, [special providence] refers to God's ongoing action, *and interaction* within and throughout human history, as well as God's actions in the personal lives of individual believers, salvific, or otherwise), as well as insights into divine inspiration, revelation, transcendence and immanence, and even Tripartitism. This same dualism also resolves a host of hitherto insurmountable theological and scientific conundrums that have long plagued Judeo-Christianity.

Such an unusually high degree of interdisciplinary explanatory power is unprecedented and may hint at a potential ontic unification between science and theology (or more accurately, between *science and Christianity*). Incredibly, quantum dualism provides penetrating insights into the very mechanics of Christianity, and does so while also resolving a host of long held disparities between orthodox Christian doctrine and orthodox scientific truth.

There is an undeniable link that exists between quantum mechanics and classical Christian theism. And it is so profound that it reveals the existence of a divine Mind and a divine purpose behind both. The link between quantum mechanics and Christianity is rooted in the distinctly platonistic *mathematical realism*[23] of the Bohr/Heisenberg ontology; which identifies unmeasured quantum systems (i.e., wave-functions) as

---

[23] Mathematical realism is a doctrine which asserts that mathematical objects, entities and realities, (e.g., geometrical shapes, natural numbers, functions, sets, axioms, proofs, truths, etc.) exist objectively and are abstract, timeless, eternal. Most mathematicians are mathematical realists and thus hold to one form or other of platonism. The discovery of the existence of quantum wave-functions (as purely stochastical entities) has effectively validated mathematical realism.

abstract probability distributions that evolve stochastically within some transcendent (and *uniquely platonic*) configuration space[24] not located within the three-dimensions of ordinary Euclidean space. What this means simply is that the fundamental essence of unmeasured quantum reality is wave-function!

Again, wave-functions are abstract probability waves that evolve stochastically over time.[25] Hence as these systems propagate they also generate statistical probabilities. These probabilities represent possibilities and/or the likelihoods for various potentialities coming into being upon wave-function collapse. The probabilities then are mathematical probabilities, and are assigned unto the various "potentialities" present at any given moment within the stochastically evolving wave-function—*potentialities* that may or may not become physically actualized upon the collapse of the system—depending of course upon the odds for each potentiality. (Formally, these statistically-based, mathematically-derived potentialities represent solutions to the famed Schrödinger wave-equation).

As mentioned above, the shift from quantum potentiality to physical actuality occurs the instant a wave-function interacts in a marked[26] way with physical reality. This interaction triggers instantaneous wave-function collapse, and the instantaneous appearance of a physical particle or particle-system. What's more, the instantaneously actuated outcome (though statistically constrained) is *indeterministic!* At least it appears that way to science. As such, there is nothing in the physical universe that determines the outcomes of quantum events! They are "naturally" undetermined, and/or *indeterministic!* Ergo the "quantum uncertainty principle."

---

[24] This space, dubbed "Hilbert space" (after David Hilbert) by mathematicians is a purely abstract space of conceptual and/or logical possibilities.

[25] While time dependent wave-functions can and do stochastically evolve in a classical temporal, deterministic manner, their associated quantum event outcomes manifest indeterministically upon wave-function collapse. Yet both indeterministic wave-function collapse and deterministic wave-function evolution transcend ordinary spacetime, taking place in some or other realm of platonic configuration space. (See Wheeler's delayed choice experiment [1978]). So then, while quantum wave-functions may dynamically evolve in a classical, temporal (i.e., deterministic) manner (again, stochastically and/or mathematically), the systems in their entirety are beyond the constraints of ordinary spatiotemporal reality. See *Mathematical Foundations of Quantum Theory*, edited by A.R. Marlow, Academic Press, 1978.

[26] i.e., irreversibly.

This gives way to a paradox of sorts. Quantum outcomes do obtain, *but without sufficient causal precedent.* Quantum event outcomes are thus deemed a-causal. *Yet they do occur!* This gives rise to the question: If quantum event outcomes are not determined by anything within the universe, could quantum event outcomes be sufficiently determined by something outside of, or beyond the universe?

For, if quantum event outcomes do obtain (and we know they do because they physically actuate particular quantum states into being), yet there is nothing in the physical universe that determines their outcomes, then actuated states must be determined by something transcendent to, and wholly beyond this physical universe?

Recall that quantum wave-functions are both wholly ethereal and transcendent. As such, for any 'determining agent' to interface with wave-function, it too would have to be wholly ethereal and transcendent—*and necessarily so.*

Put differently: if the physical universe is "all there is" and there is nothing within the physical universe that determines quantum event outcomes; yet quantum event outcomes *are* determined, then the universe obviously isn't "all there is", and quantum event outcomes must obtain their determinations from beyond the universe. This in turn means that there is something beyond the universe capable of preferentially choosing particular quantum outcomes to the exclusion of all others. This is most telling. Wave-functions after all are abstract, mentalistic, and platonic, such that our determining agent (or rather, Agent) must not only be transcendent, but also disembodied, as well as freely willed!

The picture we are thus left with is that of a physical universe emanating into being (constituent-particle-by-constituent-particle) from an utterly abstract, transpatiotemporal medium (the quantum void) that is strikingly platonic. Contra material reductionist claims that all (including mind) comes from matter, it appears that the mirror opposite is the case.

As far as this physical universe is concerned, all is being continuously *willed* into physical existence (again, constituent-particle-by-constituent-particle) by what seems to be a transcendent volition—a preferential 'Will' so to speak. Whatever the ethereal quantum wave-function is, it is a perfect medium for allowing the disembodied Mind (or Spirit) that is God to interface with, and indeed *manifest,* the physical world in a *creation continua* style manner.[27]

Summing over, 'quantum actualizations' occur the instant a wave-function interacts (irreversibly) with a macro-matter-aggregate of physical reality. What's more, the particularity of these actualized outcomes are entirely indeterministic. And it is this indeterminism and/or a-causality that suggests a transcendent Determiner of quantum events. Determinations of this sort, that is, determinations induced by a transcendent mind (or Mind) would necessarily be volitionally induced. This is very theistically suggestive, for the picture that emerges is that of a physical universe emanating forth from an otherworldly, transpatiotemporal (platonic-like) medium, being continuously *willed* into physical existence, particle-by-particle, by what seems to be a transcendent volition—or "Determiner" of quantum outcomes. As such, "whatever" the ethereal wave-function is, it is a perfect medium for allowing the disembodied Mind of God to interface with the physical world.[28]

The dualism of quantum reality is an ontological dualism consisting of (1) incorporeal wave-function, and (2) particulate matter.[29] This matter/wave-function dualism is distinctly reminiscent of the classical mind/matter dualism of Descartes. Technically however, we are not dealing with two distinct substances, i.e., one of mind, and one of matter; we are dealing with a solitary substance capable of manifesting

---

[27] Quantum event actuation speaks to the doctrine of *creation continua* and hence the existence of an intrinsically "theistic" Deity. Unlike the god of 'deism' who creates the world, then, afterward recedes into oblivion never to be seen or heard from again, the 'quantum Creator' is a theistic deity who continues in the act of creation, even after the initial creation event (for instance, determining particulate reality into being one actuation at a time).

[28] Not unlike the apophatic Trinitarian God of Judeo-Christendom, the wave-function is an objectively existent reality that is conceptually inexpressible and/or beyond the cognitive dexterity of the human mind to grasp. Yet, as quantum physics has demonstrated, such "*meta*-realities" can *and do* exist *"objectively."* And while such realities may defy pictorial visualization and/or representation, the human mind is not so rigid that it cannot "know of" such things through less direct avenues, such as abstract mathematics.

[29] In 1905 Einstein demonstrated the particulate nature of light waves via his famed "photoelectric effect" for which he won the 1921 Nobel Prize. Inspired by Einstein's discovery, a French Prince by the name of Louise de Broglie pondered whether matter particles might not in turn display wave characteristics! In 1923 de Broglie set forth his "matter-wave" hypothesis, which was subsequently proven in 1927 when Davisson and Germer demonstrated the diffraction of electrons in mercury vapor.

dual properties in an either-or like manner; one wave-like (and/or mind-like) and the other particle-like (and/or matter-like).

Either way, there are two distinct substances that manifest over/and/against each other in a highly dualistic manner. I thus speak of wave-particle duality in terms of a dualism; quantum waves and quantum particles are, after all, dueling properties in the sense that each can only ever exist at the ontological expense of the other.

This unique brand of quantum dualism turns out to be ontologically necessary for mind-to-brain (or soul-to-brain) interfacing to occur.[30] As

---

[30] The brain is in many ways the central mainframe, or mission control center of the body, and there is compelling evidence to suggest that spiritual (and/or metaphysical) elements exist which govern the workings of the brain, not unlike Ryle's "ghost in the machine." The 'Rylean ghost' of course, is none other than the "soul" of classical Christianity and its means and/or medium of soul-to-brain interfaction is wave-function. As we have seen, quantum indeterminacy provides God with the means to interact within nature in a manner that is both naturally and scientifically lawful, and hence theologically unproblematic. Remarkably, these same indeterministic opening's, present throughout nature, and which allow God to act in the world, are also present at critical loci within the brain (namely, the synaptic junctures) allowing for the potential influence of mind (or soul) upon matter (that is, grey matter)! At its lowest levels, the brain is a quantum system that is uniquely suited for soul-to-brain interfaction. This is possible due to specific molecules within the brain (such as NH^2 [Wolf, 1989]) that are seemingly distinctly suited for the mind/soul-to-body/brain conveyance of will. The implications of quantum mechanics (and quantum wave-function in particular) for dualist-interactionist models of mind has not been lost on neuroscientists. The late great neurophysiologist Sir John Eccles writes; "[T]he hypothesis is that mind-brain interaction is analogous to a probability field of quantum mechanics, which has neither mass nor energy yet can cause effective action at microsites. More specifically it is proposed that the mental concentration involved in intentions or planned thinking can cause neural events by a process analogous to the probability fields of quantum mechanics." (Sir John Eccles, *Evolution of the Brain: Creation of the Self* (London and New York: Routledge, 1989). Dualism, while not exactly popular at this moment in time, has in no way been threatened (much less defeated) by present-day naturalistic accounts of consciousness. (See for instance, Richard Swinburne's *Mind Brain and Free Will* [2013], and/or Alvin Plantinga's, *Where the Conflict Really Lies* [2010], and others, [see below]). Despite the overconfident assertions of materialist thinkers, all monistic accounts of consciousness fail in one way or another. Matter alone simply cannot account for personhood! Eccles therefore concludes: "Since materialist solutions fail to account for our experienced uniqueness, I am constrained to attribute the uniqueness of the Self or Soul to a supernatural spiritual creation." (ibid). Eccles is far from alone in his assertion that mentalisms are inexplicable in purely physicalistic terms, as a host of prominent figures in the mind sciences have come to similar conclusions. Notable modern-day dualist thinkers include British biologist Sir Alister Hardy, Princeton philosopher Frank Jackson, Canadian neurosurgeon Wilder Graves Penfield, Oxford philosopher and psychologist Daniel N. Robinson, and neurophysiologist Roger Wolcott Sperry, not to mention the great philosopher of science Karl Popper, who, along with quantum physicists Friedrich Beck and Henry Margenau either worked or wrote with Eccles on the development of a "dualist-interactionism" model of mind/brain interfaction, based

we will see, the key tenets and/or central phenomena of quantum theory are in one form or another ontologically necessary for the realization of Christian theism. This is an incredible discovery and provides a considerable hint at the true nature of ultimate reality.[31]

Quantum Indeterminism

Another key tenet of quantum theory is "indeterminism." Quantum indeterminacy is rooted in wave-function collapse (which is just one aspect of quantum process). You see, prior to collapse, wave-functions evolve in a rigidly deterministic manner, giving rise to potentialities in strict accordance with the Schrodinger wave-equation. However, when a wave-function encounters a macroscopic system such as a measuring device, the evolving wave-function instantaneously vanishes, collapsing away to nothingness.[32]

At this self-same instant every potentiality of the abstractly evolving wave-function vanishes; that is, except for one—the one that becomes physically actualized in space and time. And it is this outcome that is indeterministically realized.[33] That is to say, quantum-event-outcomes are insufficiently determined by causal antecedents—for, according to quantum orthodoxy, these have no causal antecedents! As such, there is an implicit uncertainty associated with every single quantum event outcome. And it is here, within this miniscule window of ontic uncertainty that the will of God is exacted upon matter, and hence,

---

upon Eccles' having successfully isolated a quantum mechanical interface wherein mind is capable of influencing microsites within the synapse of the brain. Contra the materialmonist mantra that dualism is dead, dualism is very much alive and well within academia among leading thinkers and neuroscientists, and is more than satisfactorily defended in a number of excellent works by authors such as John Foster, David Hodgson, and David Chalmers, and others.

[31] This ultimate reality is neither reductionistic, nor deterministic, nor materially monistic, but is rather holistic (via quantum non-locality and/or entanglement), indeterministic (via quantum a-causality), and mentalistic (via quantum wave-function). And (as alluded to) these latter three precepts, as they appear together in nature, are actually ontologically constitutive to a classical Christian cosmos.

[32] Whereas wave-function evolution is deterministic and time-reversible, wave-function collapse is indeterministic and time-irreversible.

[33] Quantum indeterminism is a consequence of quantum-ontological-dualism; consider for instance the ontic impossibility of simultaneously measuring the exact values of complementary conjugate variables (e.g., position/momentum etc.).

within the world. It would thus seem that quantum indeterminacy serves as the cloak of divine action.

Quantum indeterminacy then, refers to the uncertainty implicit within quantum event outcomes. Metaphysically speaking, much takes place within the "instant" that is the quantum event. As we have seen, during this instant the wave-function collapses, marking the physical actuation of a quantum particle or system, such that what was a moment before merely an abstract potentiality, instantaneously shifts to a physical particulate state.[34] The quantum event is no less than an ontic shift between dualist substances—a shift marked by indeterminacy. This shift, from the overtly spiritual substance of wave-function to the physical substance of materiality, is, as we shall see, a phenomenon that is both requisite and basic to the mechanics of Christianity.

It is important to realize that the indeterminacy implicit within quantum systems is ontological rather than epistemological.[35] Consequently, quantum events will always possess an intrinsic degree of uncertainty and there is nothing that science will ever be able to do to remedy the situation. There will therefore always be a gap in human knowledge owing to the ontological gap in quantum reality. Absolute knowledge then, is simply not an option at the quantum level.

There is simply no way to predict with 100% accuracy which wave-function potentialities will be physically actualized in spacetime, and which will collapse away into nothingness. The reason being, that quantum event outcomes simply have no physical determining causes, such that there is no antecedent causal chain to follow either backward in time or forward in time as was the case in Newtonian mechanics. The problem, then, is quantum a-causality, and it underwrites every single particle in the universe.

Whereas the Newtonian universe was rigidly causal and hence deterministic, the quantum universe is rigidly a-causal and hence indeterministic. The cosmic situation, then, is this: The universe is constituted of quantum particles endowed with the capacity to instantaneously (and indeterministically) actuate in and out of existence as they shift from abstract wave-function to particulate matter and back

---

[34] By "quantum system" I mean simply a multi-particulate arrangement (such as an atom or molecule) with overlapping/entangled wave-functions.
[35] *Epistemology* is the study of "how we know what we know", and concerns the nature of knowledge acquisition, methodology, scope, limitations, etc.

again.³⁶ This is the situation, and its implications for reality are more than a little profound. For, here are physical creations *without* physical causes! Here, are physical determinates *without* physical determinations!³⁷

This however, is not to say there is not a "non-physical" determiner (or rather, "Determiner") that is somehow involved, as I most certainly believe there is. After all, quantum theory only tells us that there is no "physical" and/or "natural" cause or determiner of quantum events. What's more, we know that since particulate matter actuates into being from an abstractive essence that is objectively real, that if such events do indeed have a cause or determiner, it will itself be transcendent and incorporeal. Moreover, if this transcendent incorporeality is apt to make particularized determinations it must possess intention, preference, and/or will (Or should I say "Will"). Accordingly, if the wave-function is anything, it is a sphere of volition. A realm of existential potentiality. And behind this quantum cloak of uncertainty lurks the divine Determiner—the Mind of God, freely willing reality into being one particle at a time—undoubtedly, as seen fit from the divine perspective, and in accordance with the divine will.

Quantum A-causality

> *The invalidity of the law of causality is definitively proved by quantum mechanics.*
> —Werner Heisenberg

Quantum indeterminacy is of profound metaphysical significance in that it manifests as a-causality. For, where there is "a-causality" there is mystery in that you have specific entities with specific states manifesting outright in the absence of any antecedent causal source.

The locus of quantum a-causality (and thus, the locus of 'mystery') is at, or within, the discontinuous collapse of the quantum wave-function. For, at or within this a-temporal instant there occurs the indeterministic actualization of the quantum event—which is an instantaneous physical

---

[36] Even this 'physicality' however, will always and necessarily be 'in part' as is reflected in the indeterminacy of simultaneous properties (position, momentum, etc.). Quantum reality is so profoundly and fundamentally metaphysical that even its physical ontology is tenuous.

[37] Going beyond the empirical appearance of quantum a-causality we can infer the existence of metaphysical causes and metaphysical determinations.

manifestation of a particular quantum particle or system in a particular quantum state. One instant there is nothing more than an abstract mathematical formalism (or construct) expressing what are merely the potentially-existent-states that a quantum system may or may not manifest upon wave-function collapse, and the next instant there is a physical quantum object, actualized as it were, out of thin air, and realizing one of the heretofore noumenal[38] quantum-state-potentialities.[39]

This is truly an incredible state of affairs, insofar as it reveals that the whole of physical reality emerges from an abstract construct of mathematical rationality, via the determining will of a principally unseen, undetectable, transcendent Mind.

At the end of the day, the simple yet remarkable truth is that quantum events are uncaused—*yet they occur!*[40] This is truly bizarre. Our minds are simply not accustomed to such phenomenon. After all, we live at the ordinary, everyday level of classical macroscopic phenomena, where the laws of cause and effect govern unfettered.[41] And for this reason we are tempted ask; "but what *really* determines a quantum event?" However, the answer is still "nothing". Regardless of how we may feel about it, quantum events are a-causal—at least, according to the physicalistic rules of naturalistic science.

Yet, even so, one cannot shake the sentiment that something more interesting is going on the instant of a quantum event. After all, as theologically minded individuals know, only God is truly without a cause.[42] This is easy enough to comprehend, insofar as God in His transcendence is beyond or outside of the causal spacetime continuum of the universe. Quanta, on the other hand, are part of this physical reality—hand, are part of this physical reality or, at least they are *half of the time*.[43]

---

[38] The term "noumenal" here and elsewhere should not be taken as referring to any strain of Kantian idealism. Rather it is meant simply to convey the pure ethereality and mentalistic attributes of quantum wave-function.

[39] These "potentialities" are in fact solutions to the Schrödinger wave-equation.

[40] That is, quantum events are "uncaused" from a naturalistic, scientific perspective. In other words, there is nothing within this physical universe that determines quantum event outcomes!

[41] Even chaotic events are ultimately causally determined.

[42] See the Kalām cosmological argument. e.g., Craig, *The Kalām Cosmological Argument*, (Eugene, OR, Wipf and Stock, 2000).

[43] That is, after wave-function collapse.

Recall quantum dualism, and how physical reality actualizes out from the quantum ethereality. Here, at the universes bottommost foundation, where the line between physics and metaphysics blurs, and where materiality condenses out of the mind-like medium of wave-function, it might not be all that absurd to imagine God acting from perhaps just beneath or behind the wave-function, to determine quantum indeterminacies.

How incredibly to discover that physical reality reduces to a preternatural substrate that is a uniquely platonic province, yet which exhibits mind-like behavior! Namely: volition, freedom of will and/or action. Such is the nature of the wave-function. Such is the nature of the divine. For, by both, mankind is put into contact with the mysterious, the eternal, the infinite—albeit in different ways, of course.[44]

Beyond Newton

If quantum physics teaches us anything at all it teaches us that the three pillars of atheistic scientism; materialism, reductionism and determinism, are dead. After three centuries of paradigmatic rule, the mechanistic reign of Newtonian science has come to its end, having been overthrown by the new quantum paradigm, which is ontologically antithetical to its predecessor paradigm, being immaterialistic, holistic, and indeterministic! What's more, it may be reasoned that if the now rescinded Newtonian paradigm had atheistic implications (which it most certainly did), then the new scientific paradigm, being its mirror opposite, should have theistic implications (which it most certainly does!).

The bottommost layer of physical reality has now been plumbed, and the ontology thereof begins and ends with that ethereal substance we call 'quantum wave-function.' This truth, if sufficiently appreciated would alter the very soul of secular science, which, for reasons having nothing to do with objective empirical fact and everything to do with subjective metaphysical bias, remains stubbornly mechanistic, reductionistic, materialistic, and ultimately atheistic.

Whereas the classical mechanist paradigm was founded upon atomism, the new quantum paradigm, being more foundational still,

---

[44] For instance, the infinity that is God is also personally knowable. Furthermore, I am not here suggesting that wave-function is itself divine; only that wave-function is a *medium* of the divine.

is founded upon that from which the atomic realm emanates; namely, wave-function and/or the ethereal quantum reality. Here, we come into contact with the underwriting metaphysical and/or spiritualistic reality from whence the ordinary everyday reality of mere physical appearance derives.

Here, we are put into contact (if, only indirectly) with the ultimate reality that lurks just beneath and/or just behind the physical cosmos. And this ultimate reality, being far from material, is not only ethereal but quite possibly spiritual—and *instrumentally* so, serving (quite conceivably) as a medium between this physical universe and the sacred reality (and/or Divine Mind) that lay just beyond it.

Quantum reality, as conceived here, serves not just as a metaphysical bridge between worlds, but also as an *instrument*. An instrument of choice. Or, should I say, an instrument "for" choice, to allow the divine Mind to make physical determinations within this material universe from a transcendent location. Recall that wave-functions are abstract, albeit objectively existent probability densities that collectively underwrite the whole of physical reality as a vast set of possible existents. The very foundation of physical reality then, is no less than a sphere of volitional choice[45], such that the whole of physical reality reduces to an intrinsically willful noumenality that is strikingly akin to the *Logos* of ancient Stoicism and early Christian thought.[46]

Unto the former, the *Logos* was the Soul of the universe; that underwriting principle of reason and rationality that governed the cosmos. It was an active, rational, generative essence closely associated with the Mind of God. Unto the latter, the *Logos* was (and is) Christ—*the Word*, who, being the same with God[47], is the rational creative Mind that spoke the cosmos into being—the very same cosmos He ultimately saves and redeems.

It thus appears that we live in a physical universe whose material constituents are quite literally 'ensouled'. Understand this notion of ensouled matter, along with the related notion of a 'world-soul' existing

---

[45] These "volitional choices" are in reality 'existential possibilities' and/or 'potential existents', and thus serve as grist for the mill of divine Will.

[46] This similarity exists only to the extent that God makes determinations by and through the "seemingly noumenal essence" of wave-function.

[47] See the Gospel of John 1:1-14 for the definitive Trinitarian statement.

beneath the physical appearance of things is no mere dreamy return to the metaphysics of ancient Greece. Rather, it is hardcore, up-to-the-minute, physics—quantum physics. And it is calling into question everything we thought we knew about reality.

As we shall see, the implications of the quantum revolution extend beyond mere naturalistic science, encompassing both the physical and the metaphysical in a seeming "theory-of-*existence*" that once and for all reconciles science and God—or more specifically, *the laws of nature,* and *the doctrines of Christianity.*

A physical/metaphysical unification of this sort is only possible because the quantum realm extends just beyond the physical limitations of the material universe, in seeming allowance for an overlapping of ontological domains. Ironically, the bottommost layer of physical reality is not even physical—it's metaphysical! And it is here, within this bottommost layer of reality that we discover the mechanics necessary for the spiritual-to-physical give-and-take of a Christian realist reality.[48]

In this work I will endeavor to demonstrate how quantum mechanics underwrites key Christian doctrines and their associated theological phenomenon.

We will for instance look at *divine action* (against the enduring questions: how can God act within a law-bound world? And, how can God, as a Spirit, possibly interact within a physical world. Both are longtime conundrums within classical theology over/against classical physics).

We will also look at the phenomenon of *human free will* (another longtime conundrum within classical theology. Both with respect to the omnipotent sovereignty of God, and with respect to soul/brain interfacing and the associated question of how the spiritual can possibly effect the physical?). Not to mention a host of primary and secondary questions, issues, paradoxes and conundrums relevant to the science/Christianity controversy—along with their quantum-based resolutions. Quantum mechanics does all this and more, and I will attempt to explain both how, and why this is so.

Having now given what barely counts as a cursory response to the earlier questions: *"what exactly is quantum mechanics?"* and *"how can*

---

[48] The term "Christian realism" used throughout this work, has nothing to do with the mid-twentieth century 'Christian realism' of Reinhold Niebuhr. I use the term simply as an expression of the objective truth and reality of the Christian God and the Christian Faith.

it possibly relate to God and Christianity?", I now move to consider a similar set of questions—these with regard to the 'divine' aspects of our inquiry.[49]

These questions are:

- *What exactly is the Christian Faith?*
- *And, how can the spiritual workings thereof possibly be realized in a physical universe?*

Brief

Critics of Christianity, and others outside the faith, tend to view Christianity as simply one religion among the myriad of world religions. However, those of us within the faith, understand Christianity to be less of a 'religion' and more of a message—a message of hope centered on God's infinite love for the world.

Among other things, the Christian message of hope is no less than an answer to that most fundamental question of human existence: "If a man die, shall he live again?" This question appears in the Book of Job, the oldest Book of the Old Testament, but it isn't satisfactorily answered until the New Testament, where the question is once again taken up, and ultimately answered, in, by, and through Jesus of Nazareth—the Christ—Savior, and ultimate hope of the world.

I believe that "hope" is the most wonderful thing in the universe. Without it, all is lost. But with it, all is possible. For, where there is hope, there is possibility; possibility of the impossible; probability of the improbable. And no individual hope burns brighter within the innermost constitution of our being, than the hope of transcending death, to be one day reunited with those loved ones who have departed before us—if only momentarily.

This hope is greater, deeper, and more profound than any mere projection.[50] For, it is "more real" than any earthly longing. It is an

---

[49] A consideration of these preliminary questions is necessary if we are to understand the nest of issues associated with the science/God controversy.
[50] The outdated projection theory of Freud, which posits religious hope as mere wishful thinking, is in no way inconsistent with the reality of the mystery of God. Projections can just as easily be understood as the human imaginations way of grappling with the ineffable

elusive, yet vividly genuine yearning for the recovery of something seemingly once lost to us. And this yearning—this hope, is symptomatic of our species. This hope is the Christian hope, and it is basic to who and what we are, both individually as persons, and collectively as a people. Thus, we are *all* familiar with the Christian hope, whether Christian or not. For, this hope is the greatest hope. And it is the hope that we all share.

This shared hope of man[51] is neither a psychological glitch, nor an unintended quirk of evolutionary history. Rather, it is an intimation of the ultimately transcendent nature of man.

Every person, including every atheist shares this hope for "something more"—something beyond the grave—regardless of what he or she may say when in like company, or when not entirely alone with themselves, such as in the dead of night when both the darkness of solitude and the lucidity of finitude loom large; when the self is all alone with itself, and that elusive "something more" that exists just beyond our grasp is most profoundly felt. During these moments *we all believe*, for we have been confronted by "something more." And this 'something more' is at once overwhelmingly intimate *and infinite*.[52] In these moments, our existential independence is also our dread, such that the only remedy is surrender.[53] For, with dependence there also comes hope. This hope is universal. And it is realized through the bodily resurrection of Jesus Christ.

Reason for the Hope within us

So, is there any objective evidence to support this universal subjective suspicion of transcendent reality? Well, besides the incontrovertible historical evidence that Jesus Christ walked this earth, and was in fact who He said He was (which evidence, no reasonable person would doubt

---

mystery and inexhaustible reality that is God. After all, theologians have long known that imagination plays a constructive role in humankind's conceptualization of God.

[51] I use the words "man" and "mankind" throughout this work out of mere convention, and surely not gender bias.

[52] Haught, *What is God, How to Think about the Divine*, (Paulist Press, Mahwah: New Jersey, 1986).

[53] Ibid.,

apart from metaphysical bias[54]), there is indeed objective evidence. And it comes from quantum physics.

Quantum physics has revealed a reality that is far more in line with the metaphysical reality depicted in Judeo-Christian Scripture than the ultra-mundane reality set forth by atheistic materialists. What's more, quantum reality not only *corroborates* Christian reality, it also *actuates* Christian reality—and does so in such a profound manner, and to such a robust degree that one may reasonably presume that quantum process itself has actually been designed to pander to Christian theism and/or the tenets thereof.

My assertion that the Christian Deity created this universe is indeed extraordinary. And if extraordinary assertions require extraordinary evidence, then the extraordinary reality of a quantum/Christian consonance (wherein the phenomenological tenets of the former are suited to both underwrite and realize the theological tenets of the latter) should more than suffice.

Of the tenets of Christianity, 'divine action' is perhaps the most fundamental. For, the hope of Christianity, which is the hope of mankind, stands or falls on the concept of a personal God—and thus, on the notion of a God who acts. And not only acts, but *has acted* in both history and the world.

Traditional theologians generally identify three distinct forms of divine action. There is miraculous divine action, wherein God brings to pass that which is seemingly contrary to nature. There is special divine action, wherein God acts agentially (that is, causally) in nature, within the pre-existing causal structures of nature's laws and processes. And there is general divine action, wherein God acts by and through the secondary processes of nature (of which God is the primary cause). (Note that special divine action differs from miraculous divine action in that the former doesn't necessarily entail the violation and/or suspension of scientific law. Rather, God acts agentially in the world, inside the limits of natural law. Special divine action then, is God acting in the

---

[54] I speak here of the Schleiermachian inspired assumption of naturalism (and/or bias against supernaturalism) that so heavily influenced 19th and twentieth century Biblical criticism. For the definitive critique of liberal theology see Karl Barth's *Protestant Theology in the Nineteenth Century* (1946); Barth's *The Epistle to the Romans* (1922); and Barth's *Church Dogmatics* (1932-1967). For a scholarly account of the resurrection see also Barth, as well as N.T. Wrights *The Resurrection of the Son of God; Christian Origins and the Question of God, Vol.3.* (2003).

world as an 'Agent among agents' or 'Cause among causes). An example of MDA would be Jesus walking on the Sea of Galilee (Matt 14: 25). An example of SDA would be God bringing forth the strong east wind that divided the waters of the red sea (Ex 14:21). And an example of GDA would be God bringing forth the spring rains in their season (Zech 10:1).

The personal nature of God is further attested unto in the witness of subjective personal experience. For, even when God isn't acting directly in nature and the world, the *senses divinitatis* (or sense of divinity) remains a compelling witness unto mankind of the existence of the transcendent—the infinite—the eternal. Hence, the universal, albeit inexplicable religiosity of man—both across the globe and throughout history. God is both present and active in the world.

A God that is not personal (such as the Newtonian inspired god of deism[55]) cares neither for His creation, nor for its creatures. Such a god creates, and then fades into oblivion. Such as god is useless, except possibly in a "gaps" scenario, as an explanation for the world's origin. The Christian God, however, being the true Creator of the world is concerned not only for its origin, but also for its end. And the story of this worlds end is *redemption* and *renewal*—both for creature *and* creation. For, this world, as creation, is also the stage of a divine drama; a world brought into being by the will of the Father, but bought for eternity by the obedience of the Son.

We thus see *just how fundamental* divine action and the notion of a personally active, directly involved God *is* to the Christian drama. And here is where quantum mechanics comes in. There exists a quite clever, and quite *coincidental* correlation between quantum mechanics

---

[55] Because Newtonian physics was atomistic and deterministic it was interpreted as being also reductionistic and materialistic. Enlightenment thinkers such as Hobbes, Voltaire, Laplace, the French Philosophes, et al., understood Newtonian mechanics as evidencing a causally-closed mechanistic world-system that was either deistic or atheistic in that it left no causal space within nature for either divine interaction of soul-body interfaction. Ironically, contemporary materialist thought is grounded on the self-same (Newtonian-inspired) metaphysics as was Enlightenment skepticism—despite the fact that quantum theory has quashed the Newtonian metaphysic. Early-modern materialist works include Hobbes' *Leviathan* (1651), Voltaire's *Elémens de la philosophie de Neuton* or, *Elements of Newtonian Physics* (1738), Laplace's *Mécanique Céleste* or, *Celestial Mechanics* (1799-1825), and Julien Offroy de La Mettrie's controversial (1747) work *Homme Machine* (or, *Man a Machine*), to name just a few. Noted French atheists and deists of the period include Claude-Adrien Helvetius (1715-1771), Jean-Jacques Rousseau (1712-1778), Denis Diderot (1713-1784), and Paul-Henri-Dietrich d'Holbach (1723-1789), again, to name just a few.

and the type of mechanics necessary for divine action to occur—both in nature, and in a seemingly naturalistic way that is objective, but not contrary to the natural course of causal regularity. But this is just one of many such "clever correlates" wherein the relative fitness of quantum phenomenology appears ideally suited to realizing the often metaphysically-based phenomenology of Christian doctrine.

If my suspicions are correct regarding these "ideally suited" phenomena, then a disembodied (albeit personal) 'Intelligence' created a physical universe underwritten by a very particular set of quasi-physical processes for to allow itself (*as a disembodied agent*) a particular degree of freedom-of-action in nature—albeit, cleverly cloaked in the causal anonymity afforded by quantum process.

This is no mean engineering feat. In fact, I believe it to be the most elegant and most ingenious engineering feat ever conceived. Seemingly, against all reason, God managed to create a physical universe governed by physical laws, yet which at the same time panders to spiritual phenomenology! And a very particularized spiritual phenomenology at that; namely, that of the Christian faith.

What is most impressive here is that God has managed to cross-ontologize spiritual reality and physical reality. What's more, he has set it up in such a way that the reality of God will exist 'overtly' unto believers, but 'covertly' unto the naysayers and secular masses (as well as within the world's physics laboratories). Anonymity of this sort is absolutely necessary in that Christian theology dictates that salvation be uncoerced. Salvation after all, is not an intellectual exercise (as contemporary atheists attempt to make it). Rather, it is a personal encounter: it is allowing oneself to be grasped by that which cannot itself be grasped. Salvation is a personal gift, and as such, must be *freely willed* unto personal acceptance. It is a One-to-one encounter with the living God.

The fitness of quantum phenomenology for Christian theology is most impressive. So elegant and so economical are the quantum-ontological solutions to Christian metaphysics that I seriously doubt they could've obtained apart from the divine Mind. In fact, so elegant, so purposive, and so particularized is the ontic utility involved that it seems it could have only been conceived by the divine Mind. The ontological architecture that lay just beneath the physicality of the cosmos is simply too robust and too specialized toward a particular type of reality to be the product of chance.

Quantum mechanics remedies the physically implausible engineering demands intrinsic to a realist Christian reality with an almost miraculous degree of utility. It appears that at the moment of creation, God established a precision network of quantum laws and principles, to, among other things, one day facilitate the divinely conceived plan of God for this creation.

We live in a universe that appears to have been purposely engineered to accommodate a very particular set of interrelated doctrines. These doctrines in turn necessitate an equally particular set of interrelated phenomena. And here is where quantum process comes in—particularly quantum indeterminism and quantum dualism. For, without these phenomenon we truly would live in the one dimensional, deterministic, materialistic, *atheistic* universe imagined by Laplace and his materialist counterparts.

Consider, any overtly materialistic universe (such as ours) needing to accommodate genuinely spiritual phenomenon (such as ours) would of necessity need to be underwritten by some or other form of a-causal, a-physical phenomenon, closely (if not identically) akin to quantum reality—that is, if it hopes to be remotely intelligible.

Having said this, quantum process may be the only way (or at least the most sensible way) for a material world to phenomenologically satisfy the preternatural demands of Christian doctrine, while simultaneously avoiding the tensions of classical dualism.

Adding credence to my assertion that quantum mechanics has been divinely purposed is the fact that the self-same quantum processes that so aptly facilitate Christian doctrine also resolve a host of paradoxes pertaining to Christian theology, both with regard to classical theology and the modern science/theology conflict (but more on this later).

That so much can be accomplished with so little (namely, the near infinitesimal value of the key quantum constant dubbed the Planck constant) speaks volumes unto the reality of both the Creator and His purposes. And these purposes, as hinted unto by quantum physics, are those of Christian *theism*—with a particular emphasis here on 'theism' in that it speaks unto the personal nature of God and the reality of His action in nature, history, and the world.

Hence, it is to the topic of divine action that we now turn.

Chapter Two

The Province of Providence

> Men cast lots to learn God's will, but God himself determines the answer.
>
> —Proverbs 16:33

The God Who Acts?

The idea of an active, personal, God is at the very heart of Western culture. Countless adherents of the Judeo-Christian faith fervently believe that God acts in history, nature, and in their own lives (such as in response to petitionary prayer). What's more, God's actions are thought to be wholly objective, such that an event occurs that would otherwise not have occurred, had God chosen not to act.

This highly personalistic view of God comes directly from scripture which urges the faithful to beseech the Lord in even the smallest of matters. The Apostle Paul for instance, directs the newly converted gentile believers at Thessalonica to "pray without ceasing" (1 Thess 5:17), and in a later epistle to the converts at Philippi, Paul writes, "Do not be anxious about anything, but in everything, by prayer and supplication, with thanksgiving, present your requests to God. (Phil 4:6). Paul's prayer directive is not dissimilar to that of Christ's. For, in Luke 18:1 Christ teaches that 'we ought always to pray', and 'never give up hope.' Christ also assures us in the Gospels that no request is to petty to ask of the Father, who is intimately concerned with even the most trivial events of creation, whether it be the "fall of a sparrow" or the "number of hairs on our head" (Matt 10:30).

Verses such as these abound in Scripture, and so clearly support the Western Judeo-Christian tradition of asking God in prayer.

Taking Divine Action Seriously

Divine action is so fundamental a conviction among Christian believers that few probably ever pause to consider in any real depth its physical implications. What exactly do we mean when we say that God brings to pass a given physical event? What actually takes place during an act of say, special providence? We may know exactly what we mean by the term "special providence" (e.g., we mean that God acts objectively in nature by acting agentially upon the physical constituents thereof [such as by initiating a novel causal sequence into nature's pre-existing causal chains]), but the truth is that we have no idea how God goes about bringing to pass an act of special providence.

In considering such things we are dealing with the 'causal mechanics' of divine action and the question of what physically and/or scientifically takes place when God (who is Spirit) interacts with the material world? This is a truly fascinating topic, yet divine action and the causal mechanics thereof, are rarely considered phenomenologically and/or scientifically, even though God's actions are believed to be wholly objective.

The question of divine spirit-to-matter interfacing is confounding to say the least and is one of the most perplexing problems of the modern science/God controversy. On this topic there are many questions and ideas to be considered. One such question/idea is that of energy transference? If God truly initiates novel causal sequences into the physical order it would seem that there must be some form or other of energy transference that takes place between the divine finger (or 'will') and the physical constituents being physically acted upon (e.g., particles, atoms, molecules, etc.,). A similar question is that of "interfaction." The question of how God (who is Spirit) goes about initiating a physical causal action. The crux of the issue being God's disembodiment—and the obvious fact that for a causal initiate to trigger a physical effect, the causal initiate must itself be physical!

Moreover, this gives rise to an issue of divine concealment, in that any such "action" (or 'causal initiation') made by God in the world would necessarily be contrary to nature's regularly occurring causal sequences (if, only agentially), thus rendering God's causative actions vulnerable to scientific detection (at least in principle).

Of course, God would never allow His will to be scientifically dissected, and so we would expect (on theological grounds) to discover that God has found a way around this particular 'causal' conundrum.

But how so? How could God act causally, without actually *acting causally?*

Paradoxically, God would need to come up with a way of covertly initiating causal action, albeit without *overtly manifesting that causal action*. This much is certain. So certain in fact, that if God does genuinely exist, we could actually expect to discover in nature, possibly even in-built within the causal structure of action itself, *a causal cloak! Perhaps even one open to disembodied causal initiates!* And of course, this is exactly what we do discover. How strange, that this incredibly unique feature of reality should exist, if a theistic God does not. But I digress.

Returning to the problem at hand for a theistic deity, the trick for God would be to come up with a means whereby to physically and efficaciously *act* in the world, albeit in a way that insures that this (necessarily metaphysical) 'means of acting' remains under the radar of science.

The solution of course was quantum mechanics. And a more elegant and economical solution I cannot imagine. For, it kills a number of theologically-problematic birds with a single stone. "Birds" which have long nested in the branches of classical divine action. By "classical" here, I simply mean those accounts of divine action formulated prior to the quantum revolution—and thus conceived against the backdrop of Newtonian determinism.[56]

---

[56] Unlike Newton (1642-1727), who posited that God intervened in nature from time to time to sure-up the stability of the solar system, Laplace interpreted Newton's deterministic laws of mechanics as suggesting a closed deterministic universe. Newtonian mechanics, insofar as it was intrinsically deterministic, seemed to also be intrinsically deistic. Deism after all imagines a distant clockmaker/winder god who after initially setting things up, ceases from any and all further engagement with the world. The God of Christianity however, both creates the world and subsequently acts within the world. This claim however, didn't exactly bode well for Christianity, set, as it was, against the Newtonian backdrop of a deterministic ontology. The problem you see, is that when it comes to divine action, ontological determinism suggests—indeed necessitates—a very particular

brand of providence; namely, predestination! (i.e., divine predetermination). You see, a deterministic economy like Newton's leaves no causal space within creation for divine action to occur! For, every ontological opening (so to speak) has already been causally spoken for! Consequently, the only genuine alternative of action left for God is "intervention!" That is, either agentially *intervening* in nature's otherwise closed causal economy, or else supernaturally abrogating nature's rigidly deterministic laws. In a deterministic economy then, divine intervention begs the question: why is God causally intervening in nature at all, when, from the first, He divinely predetermined all things to temporally unfold in accordance with the divine will. So you see, in a deterministic (Newtonian) cosmos there would simply be no need for God to act either specially or particularly (much less interventionistically) at any time subsequent to the initial creation event. In a deterministic universe then, there are no causal gaps, and thus no causal space for divine action; apart that is, from divine "intervention", which, as we have just seen, makes little theological sense. After all, for God to act interventionistically in a preordained economy would result in a direct clash of the divine wills. Leibniz (Newton's lifelong nemesis) criticized Newton harshly for his assertion of intervention (in light of the implications thereof). Yet Newton remained steadfast in his views, seemingly blissfully unconcerned with theological consequence. For instance, in Query 31 of his work *Optiks* (1706), Newton first argues from the elegance of the solar system to the necessity of divine design, and then from the inelegance of the solar system to the necessity of divine intervention! "For while comets move in very eccentric orbs in all manner of positions, blind fate could never make all the planets move one and the same way in orbs concentric, some inconsiderable irregularities excepted which may have arisen from the mutual actions of comets and planets on one another, and which will be apt to increase, till this system wants a reformation." Leibniz (himself a theodicean philosopher of the highest order), undoubtedly irritated by what he took to be Newton's amateurish theologizing, wrote to a friend: "Sir Isaac Newton and his followers have also a very odd opinion concerning the work of God. According to their doctrine, God Almighty wants to wind up His watch from time to time: otherwise it would cease to move. He had not, it seems, sufficient foresight to make it a perpetual motion." (Leibniz' letter to Caroline of Ansbach, H.G. Alexander, *The Leibniz-Clarke Correspondence: Together with Extracts from Newton's Principia and Opticks*, [Manchester; Manchester University Press, 1998]). A century later, Laplace (1749-1827), a fan of Leibniz (and likeminded critic of Newton's celestial interventions) claimed to have resolved the gravitational glitch dilemma—the very same dilemma that led Newton to posit periodic miracles for celestial upkeep. Laplace could thus boast in that his model of celestial mechanics had no need of a "god-of-the-gaps" hypothesis! (It is often stated that Laplace told Napoleon that 'he had no need of a God hypotheses, when, in reality what he meant was a "god-of-the-gaps" hypothesis). It thus appears that Laplace, as a deist (and sometimes atheist) understood the implications of Newtonian determinism better than Newton. And apparently so did subsequent theological thinkers. Ergo the radical shift to deism in the late 18th and 19th century. This shift was based on (1) ontological determinism, and (2) theological consistency. We simply would not expect God (as a *wise* Creator) to create a causally-closed, deterministic (and thus predestinated) universe if His intentions were, from the first, to act intermittently in that creation! Likewise, we would not expect God (as a *wise* Creator) to create a causally-open, indeterministic universe if His intentions were, from the first, to predestine (and/or predetermine) that world according to a foreordained will! "Ontology" then, may provide theology with insight into the nature of the Creator (e.g., is he deistic? Theistic?) and/or the character of the God/world relation/interface (Does God violate the laws of nature? Or is everything predetermined, having been

Paradox in Classical Providence

Of the various difficulties associated with classical divine action, perhaps none is more bewildering than that of Spirit/matter interfaction. How, it is wondered, does God go about exerting causal influence upon the physical constituents of the world, when God is a Spirit? Let us consider.

We live in a physical world, and so any causal action precipitated by God within this world will necessitate that a metaphysical-to-physical interaction take place between the incorporeal finger of God and the corporeal constituents being acted thereupon. Divine action necessitates a point of interface—a locus of interaction where the finger of God comes into direct contact with the physical substrates of the material world.

But what would such a spiritual-to-material interface be like? How might it appear in nature? And what might its processes be like? Such an interface, though requisite, is nonetheless ontologically inconceivable, if not unfathomable. Even so, one thing is certain: any such interfacing system between the "there-and-the-here" would have to (necessarily) link-up with the naturally existing causal networks of physical reality—both causally and phenomenologically.

Such an interface would need to be physical in key aspects of its phenomenology (if only in part) for, to allow God the means to interject causal force-interactions into nature. But herein lies a problem. For, insofar as these "interjections" would be marked by seemingly *ex-nihilo* expenditures of energy, they would (in principle) constitute *detectable* physical phenomena in nature. And this doesn't bode well for the hiddenness-of-God principle requisite to human free will.[57]

---

foreordained from the first?). Such questions are of vital importance to the Christian faith. For, as the Newtonian-tale (told above) demonstrates, *the ontological standing of reality* can be potentially devastating to the Christian claim of a theistic God—as the 18th, 19th and twentieth centuries aptly chronicle. This fact of reality and history however, while undoubtedly troubling to the Christian reader, should also serve as a source of edification. For, while Christian theism necessitates a very demanding, very precise, very improbable ontology—this self-same ontology is (seemingly miraculously) elegantly and economically realized by quantum mechanics!

[57] In his classic work, *Evil and the God of Love*, (1966) John Hick, borrowing from Irenaeus, speaks of God maintaining an "epistemic distance" from humankind for to preserve human free will. More recently Paul K. Moser, in his powerful book *The Elusive God* (2008) argues that God's hiddenness is intended to foster personal, private, relational

But it gets even worse in that, for God to agentially introduce an efficient physical cause into the natural order is for God to be in violation of scientific law—specifically, the law of energy conservation—which is immutable (via general providence)! You see, according to scientific law, any change in the trajectory of a physical particle is an acceleration and thus requires an expenditure of energy. However, because this energy comes directly from God it will be unaccounted for by the universe. Hence, the violation. It thus appears that special providence requires the Immutable Lawgiver to violate the very laws He immutably upholds via general providence! The classical theological notion of divine action then, is profoundly problematic, leading not only to *scientific violation* but also *theological contradiction!*

"Interfaction" is thus the central conundrum of both classical divine action and modern divine action—at least insofar as modern mainstream science persists in the outmoded ideals of determinism and reductionism.

The classical dilemma thus remains: How is it possible for God (as Spirit) to 'physically' and/or 'causally' interact with the material substrates of the physical world—and do so without violating the universal law of energy conservation?

These challenges appear insurmountable. It is even hard to imagine what a response to these would even look like. Unless that is, one simply states that God can do whatever He wants—*including violating His own laws!*

Such a riposte, however, comes too cheap. After all, it is not a question of "if God *can*", but rather, "if God *will*." God, may be omnipotent, but there are still things that God *will not do*—such as lie, for instance (Titus 1:2). It is therefore reasonable think that God (*who is lawful*) will respect His own divine decrees—whether moral or natural. What's more, God is rational. It is therefore reasonable to think that God, who from the first set out to create a universe wherein exist beings with whom He can interact, would have equipped creation with the means whereby He could do just this—albeit without peril of law violation or theological contradiction!

Given its paradoxical nature, such a reality would be a lot to ask a lot to ask of any god—even an omnipotent Creator-God as magnificent

---

knowledge of both God and His love, quite apart from mere propositional knowledge that God exists.

as the Judeo-Christian God. And this is why the discovery of just such a "means" actually existing in nature would be so profoundly implicative of a theistic reality.

But again, what exactly would such a "means" look like? Such a question is of course impossible to answer *a priori*. However, we can speculate. God, for instance, might utilize some sort of spirit-to-matter interfacing system. Or perhaps institute some sort of metaphysical-to-physical-bridge to conjoin the realm of the incorporeal with the realm of the corporeal. Of course, the only way to know for sure is to go out into creation and look to see if such a "bridge between worlds" actually exists. Of course, if such a bridge does in fact exist in nature, there is no guarantee that we, with our finite minds would be able to identify it. However, the argument can be made that since the divine creative Mind is rational, and since we are created in the divine image, this 'bridge' would be intelligible and thus recognizable to the human mind.

Again, however, there is just no way to know *a priori*. The only genuine alternative then, is to go out into reality and scientifically search—sort of like a 'metaphysical archaeologist' or a 'forensic theologian'. The problem however, is that one would have no idea how or where to begin the search. The only alternative then, is to hope that secular science stumbles upon it for us—which is exactly what secular science did when it discovered quantum reality!

The theistic character of quantum reality is no secret to those ordained scientists and physicist-theologians that have recently stepped onto the academic stage of interdisciplinary scholarship. Leading scholars in the science/faith dialogue, such as Robert John Russell[58], Ian

---

[58] For a contemporary scholarly treatment of divine action and the sciences see the landmark volumes generated from the Divine Action Project (DAP), a series of research conferences and publications co-sponsored by the Vatican Observatory (Vatican City State) and Russell's own Center for Theology and the Natural Sciences (Berkeley, CA.). These masterful volumes (of which Russell is the general editor) include: Inaugural Volume; 1988, *Physics, Philosophy and Theology: A Common Quest for understanding*, ed., Robert John Russell, William R. Stoeger, S.J., George V. Coyne, (Inaugural vol. Published by the Vatican Observatory in commemoration of the 300th anniversary of the publication of Sir Isaac Newton's *Philosophia Naturalis Principia Mathematica*). First Volume; 1993 (revised edition 1996), *Quantum Cosmology and the Laws of Nature: Scientific Perspectives on Divine Action*, ed., Robert John Russell, Nancey Murphy, C.J. Isham. Second Volume; 1995, *Chaos and Complexity: Scientific Perspectives on Divine Action*, ed., Robert John Russell, Nancey Murphy, Arthur Peacocke. Third Volume; 1998, *Evolutionary and Molecular Biology: Scientific Perspectives on Divine Action*, ed., Robert John Russell, William R. Stoeger, S.J., Francisco Ayala. Fourth Volume; 1999, *Neuroscience and the Person: Scientific Perspectives on Divine Action*, ed., Robert John

Barbour, John Polkinghorne, Nancey Murphy, Thomas F. Tracy, et al, have all written in one way or another on the implications of quantum indeterminism for divine action within a theistic worldview.[59] The works and writings of these thinkers, while familiar among upper echelon academics concerned with the science/faith dialogue, is scarcely known to the mainstream masses concerned with the same. This is undoubtedly due to the abstract (but not necessarily technical) nature of the subject matter (which can be understood for the most part without reference to mathematics). Quantum physics after all, is much harder to believe than it is to understand. One might even say that when it comes to quantum mechanics, it is not what people don't understand that gives them pause, but what they do!

Even so, in this work I offer my own novel contributions to this exciting new field of interdisciplinary study.[60] My focus being on the ontological dualism of quantum mechanics and its implications for

---

Russell, Nancey Murphy, Theo C. Meyering, Michael A. Arbib. Fifth Volume; 2001, *Quantum Mechanics: Scientific Perspectives on Divine Action*, ed., Robert John Russell, Philip Clayton, Kirk Wegter-McNelly, John Polkinghorne. Capstone Volume: Sixth Volume; 2007, *Scientific Perspectives on Divine Action: Twenty Years of Challenge and Progress*, ed., Robert John Russell, Nancey Murphy, William R. Stoeger, S.J., (Volumes I-VI Published jointly by the Vatican Observatory [Vatican City State] and the Center for Theology and the Natural Sciences [Berkeley, CA.]).

[59] Robert John Russell is the world's leading thinker in the area of 'quantum divine action', or what Russell refers to as "non-interventionist, objective, divine action" (or "NIODA"). Russell's scholarship has been particularly invaluable to me, and has greatly assisted me in my own research. Russell's works, along with the extraordinary wealth of literature generated from the DAP conferences have been tremendously influential in my studies. However, my work differs from Russell's and others in significant ways. For instance; whereas Russell and his peers (particularly, Thomas Tracy and Nancey Murphy) tend to focus on the nature and frequency of God's role in quantum indeterminacy, my work (while also speaking to God's role in quantum indeterminacy) tends to emphasize the implications of quantum ontological dualism as a possible means and locus for divine Mind-to-world interfaction. My core assertion being that the quantum phenomena of non-locality, indeterminacy, and abstract wave-function realism allows God a quite ingenious (and quite elegant) means of simply "willing" quantum outcomes into being without either violating natural law, or rendering His existence vulnerable to scientific detection. At the root of my assertion is my identification of the quantum wave-function as a platonic medium which, as an eternal reality itself, serves as a perfect interface between God (who is Spirit) and the material world. My work then, perhaps foolishly, attempts to identify the mechanism, the means, and the manner in which God acts in the world, while also endeavoring to demonstrate how a number of key tenets of Christianity are in fact grounded in quantum phenomenology.

[60] What I posit is a quantum-based Christian understanding of reality based upon a Copanhagenist interpretation of quantum theory, and a platonist interpretation of quantum wave-function.

Christian doctrine, and reality itself. For, it seems to me that quantum reality is simply too "theism-friendly" not to have been the product of purposive design.

God's Mysterious Ways

The provident hand is profoundly inscrutable. So much so, that (as noted) believers rarely approach the concept of divine action from an academic or intellectual standpoint. Most of the time, explanations for would-be divine acts are simply glossed over with the dismissive phrase "God works in mysterious ways." This statement however, is only half right. For, while the mysterious nature of 'God's ways' with respect to *divine purpose* is hardly in dispute, this does not mean that 'God's ways' with respect to "action" and "causation" are equally beyond the limits of rational human inquiry and/or comprehension. After all, it is not Gods "purposes" that we seek to comprehend anyway (as if we could possibly know the Mind of God), but only how it is that God interfaces with the material world in order to bring those purposes to pass. Thus, it is insight into the means and mechanisms of divine action that we ultimately seek.

Insofar as divine action (or rather, divine *interaction*) is in part a physical process, and insofar as the physical world is governed by static laws and regularities, there exists the possibility for discovering (at least in part) how God interfaces with the physical world. I, for one do not believe that the *ways of God*, as they pertain to God's physical interaction with the material world, are beyond comprehension. Having said that, let me restate that while I *do* believe that we may be capable of grasping the mechanics of divine action, I *do not* believe that we can in any way know God's will with respect to divine action.

For the first time in history we possess the scientific acumen to detect the interactive sphere from whence the divine finger (or rather, divine "Mind") interfaces with the physical world. The key is quantum dualism and it is unlike anything we ever even thought to consider.

The *mechanics* of divine action is a topic as fascinating as it is novel. It is fraught with remarkable scientific subtleties and exciting theological complexities. But none of the sort to prevent us from penetrating the heart of the mystery of the mechanism of providence. For shrouded within the anonymity of quantum indeterminism is the outworking will of the Creator.

Divine Mind and DNA

Nowhere are God's 'mysterious ways' more ambiguous than with respect to creation and the nature of God's creative hand in natural history.

Various Creationists and Intelligent Design theorists have sought to quasi-scientifically (or worse pseudo-scientifically) *tease out* from nature where God imparted design—specifically biological design. These efforts however, insofar as they inevitably reduce to gap-arguments have ultimately proven futile, such that as of present, the creative/causal mechanism of God remains stubbornly elusive.

If the actionary-trigger utilized by God during the creative course of natural history were someday actually discovered, it would be a revolutionary breakthrough of unparalleled proportions.[61] In light of neo-Darwinian theory, this causal mechanism would undoubtedly be molecular and/or genotypic in nature, such that the divine impartation of biological design could be transferred via the genetic-informational-downloading-system already implicit within every living thing. (Assuming of course, that God did in fact impart biological input at critical junctures throughout the course of evolutionary history). Modern science then, actually provides us with a hint as to where we might expect to find such as mechanism—that is, if such a mechanism actually existed. Such a mechanism, if found, should be in step with what modern science has discovered with regard to the molecular basis of life and the unique nature of genetic information systems.

Of course, knowing that the biological information behind morphological design originates in the genetic code is one thing, but knowing how that biological information/design is imparted from the Mind of God to the genetic code is quite another.

We are again faced with the problem of imagining how God, who is 'Spirit' could possibly interact in a causally efficacious manner with the genetic material of DNA and mRNA. We are seemingly back

---

[61] Assuming of course, that such a trigger exists.

at the Cartesian paradox of comprehending how incorporeal mind might possibly interface with physical matter—or in this case, 'how incorporeal "Mind" might possibly interface with physical matter.'

Having now determined where in nature biological design is most likely to be imparted (i.e., within the genotype of organisms) we are left to puzzle over the mechanism and means of divine information conveyance.

Of course, it would be preposterous for us to expect the DNA molecule to have come readily equipped with a mind-to-matter interfacing system to allow the disembodied Mind of God to directly upload design-data into the genetic codes of living organisms. Admittedly, such an expectation would be both absurd and fanciful. How remarkable is it then, that when we look closely at the workings of the DNA molecule we discover just such an information/amplification system readily in place, capable as it were of downloading abstract informational uploads into what is no less than a quantum database, wherein data, once received is then biologically processed via a series of complex physiological amplification systems to be actualized phenotypically and/or morphologically.

Quantum physics has brought us closer than ever before to unlocking the great "whodunnit" of existence. Against the reality of quantum transcendence, we glimpse the far greater reality of a God who is not only there, but active as well.

Who could have imagined that a new physics would provide metaphysical insights into life's most pressing questions; questions that have long plagued our species. Questions such as where did the universe come from? Why is there something rather than nothing? Is there a God? And if so, who? What is its nature? What is its relation to us?

As stated, our quest in this work will be akin to that of a "Forensic theology", the goal being, (among other things) to identify the perpetrator responsible for, the "act of creation." And as we shall see, the "perp" in question turns out to be none other than the Trinitarian God of Judeo-Christianity!

For, this "perp" left behind a very subtle, yet very distinct series of calling cards within creation. These exist within the ontic structure of quantum reality, and can leave no doubt as to who the Creator truly is. For they reveal, once and for all, the Creator's motive for creating the universe in the first place. And this motive appears to be to *manifest a*

Christian realist reality, wherein *a created humanity can partake of, and enjoy a personal relationship with the Creator Himself!*

Let us for now set aside particular issues of biological design mechanisms and return to the broader question of how God may act causally in nature, history, and the world.

Key Quantum Precepts

Uncovering the causal initiates of the divine hand within the physical world is no mean feat. Simply determining where to look; what to look for; and how to go about looking for it, is enough to tax even the greatest of human minds. Upon reflection, the problem immediately presents itself. The problem is that we are limited to conduct our research from *within* the confines of this physical world. And from within this naturalistic bubble we must try to imagine how a disembodied, transcendent Being might go about introducing causal influences into the seemingly closed causal systems of the natural world.

In any other century this quest would be impossibly futile. And admittedly, the same would apply today were it not for recent discoveries in quantum physics pertaining to the ontologically *indeterministic*, and ontologically *dualistic* nature of sub-atomic reality. For, here, deep within the phenomenological heart of the quantum ontic resides the *secret* of the divine interface. And *this secret* we can at last begin to fathom.[62]

Prior the birth of quantum mechanics, any quest to uncover the secrets of the divine hand could have reasonably been deemed a fool's

---

[62] I should point out that while we may now be able to fathom (in part) the mystery of the mechanics of divine action, *the nature of these mechanics* precludes us from being able to 'prove' divine action in any empirical sense. I am not here speaking of the patently obvious point that we cannot know that God has acted. Rather, I am referring to the remarkable fact that the ontic structure of quantum reality not only ingeniously realizes divine action but also quite cleverly *cloaks divine action!* In this regard the ambiguity in quantum process is actually a theological necessity. Furthermore, this ambiguity appears at a key place (and in a key manner) within the quantum mechanical dynamic for to ideally conceal (and in fact assist) a disembodied, transcendent agent intent on determining quantum outcomes. If God does in fact act in the world via quantum process then ambiguity of the quantum sort is exactly what we would expect to find. So while we may be able to indirectly glimpse the mechanism of providence, we can never directly perceive providence. This of course is no surprise. God may allow science glimpses of divinity here and there, but there will be absolutely no venturing beyond the sacred veil.

errand. Discoveries made in the twentieth century regarding quantum theory and the proper ontological interpretation thereof, have drastically altered our take on the nature of reality by bringing to our attention the existence of an abstract, metaphysical sphere from which the tangible world emerges.

Understand that quantum reality, being a dualistic reality, is phenomenologically both *wholly corporeal* as well as *wholly incorporeal*; both *entirely concrete* and *entirely abstract*. And when and where quantum theory is abstract, it is wholly and entirely metaphysical, being essentially and philosophically indistinguishable from a platonic entity/reality. What's more, understand that wave-function, though literally intangible, is nonetheless objectively real and/or existent.

Even more remarkable, is the fact that this abstract quantum substrate, though uniquely mind-like in essence, can and does interact with matter in a way that produces physically discernable effects.[63] And this is entirely without precedent in the material world. That is, with the possible exception of mind/body interactionism; which I suspect is itself a product of quantum phenomenology (more on this later).

The abstract quantum realm, though detectable via its physical effects, is only comprehensible to the human mind through mathematics. Stochastic expressions allow us to glimpse the pseudo-physical dynamics of quantum entities unseen. These "quantum entities" are of course wave-functions, which, insofar as they are incorporeal and also unobservable. As Niels Bohr, the father of quantum theory famously stated, "There is no quantum world. There is only an abstract quantum mechanical description."[64] Recall however that this 'abstractness' or 'incorporeality' is only one side of the quantum coin, which, being ontologically dualistic has two distinct faces, or rather 'states.' These being either abstract quantum waves (i.e., wave-functions) or physical quantum particles (such as electrons, photons, protons, etc.).[65]

---

[63] Perhaps the clearest example of quantum wave-functions generating physically discernable effects are the famed double-slit interference patterns. Such interference patterns strongly attest to the 'incorporeal wave-function realism' set forth in this work. Incorporeal/abstract wave-function realism can also be inferred from quantum tunneling, non-local influencing, entanglement, etc.

[64] Niels Bohr, *The Philosophy of Niels Bohr*, Aage Peterson (1963).

[65] Understand that as mysterious as quantum reality may be, the mystery lay *not* in what we don't know, but rather, in what we *do know*.

What this ultimately means is that the physical particles of which all things are made (including you and I) literally cease to physically exist, if and when they leave off from interacting with the macro-matter-aggregates of ordinary reality. During these periods of physical cessation, quantum entities assume an abstract mathematical status not unlike a platonic, spiritual, or even *divine* reality. Curiously, these dual quantum states (i.e., wave/particle) are not merely ontologically distinct, they are ontologically *antithetical*.

I should point out that this is not to suggest, like the ancient Gnostics, that physical matter and the quantum spirit-substance are somehow at theological odds. (The Gnostics taught that matter was intrinsically evil, and spirit was intrinsically good). Far from it, recall that the two quantum substances, while ontologically distinct are also ontologically inseparable. Matter, insofar as it emerges from the ethereal and/or spiritual, is also dependent upon the ethereal and/or spiritual! (and vice versa)

Complementarity

In quantum physics, one can measure either the wave properties, or the particle properties of a quantum system—but never both simultaneously. Niels Bohr referred to this exclusionary principle as wave/particle 'complementarity.' Bohr first presented his theory of complementarity in 1927 at the Volta conference in Como Italy. Complementarity seems a fitting term insofar as experimental knowledge gained of say, the wave-state properties of a quantum system necessarily precludes experimental knowledge of the "complementary" particle-state properties of the self-same system (and vice versa).

This bizarre either/or phenomenology, being unprecedented in nature, was completely (and conceptually) foreign to early twentieth century physicists. The anti-classical nature of the new "quantum" mechanics inspired much anxiety within the physics circles of the day, and a systematic understanding (*via* interpretation) as to "what it all meant" was desperately sought.

The question at hand was this: was classical mechanics salvageable? Or was this strange new "quantum mechanics" here to stay? An answer was fiercely sought by the physics community—albeit, with fear and trembling. At the end of the day however, (and to the vexation of classicists) the shocking answer proved to be that Newtonian physics had indeed been superseded! (If only at the atomic realm).

What emerged was the famed Copenhagen Interpretation, so-named after Niels Bohr's physics institute in Copenhagen Denmark. The first 'ontic' insights into the new "quantum" physics came from Bohr's student, peer, and close friend Werner Heisenberg. Heisenberg, working with (and often *at odds with*) Bohr in 1926-1927 on issues pertaining to quantum indeterminacy, quantum complementarity, and quantum dualism, came to propose that physical quantum properties cease to exist in well defined (i.e., determinate) states in between acts of measurement and/or observation.[66]

This implied that (1) definite knowledge of a quantum system can only be attained *post-measurement*. And that (2) definite knowledge of a quantum system doesn't exist *prior to measurement!* For, prior to measurement, quantum reality is but an indeterminate mix of potential states representing only possibilities for existence. From these abstract clouds of "possible existents" there actuates out the fixed physical quantum states from which all material reality is ultimately (if tenuously) constituted.

According to the Heisenberg ontology, which underwrites the Copenhagen interpretation then, physical quantum properties do not exist prior to measurement, whether it be momentum, position, spin, etc. And of course, that which lacks physical properties cannot properly be said to physically exist! Quantum systems (and/or their properties) however, do exist prior to measurement, they just do so *non-physically*, that is, they exist in the abstract, as mere *potentia*—*potentia* which may or may not actuate into physicality upon measurement. It is during the process of measurementation that indeterminacy arises, as there is an implicit element of chance in the actuation of quantum properties. These indeterminacies are constrained by stochastic laws—laws which mathematically govern the abstract side of quantum reality.

---

[66] The first insight into the indeterministic nature of quantum reality came in 1927, when Werner Heisenberg published his groundbreaking "uncertainty principle" which tells us that nature places a fundamental limit on the degree of accuracy with which it is possible to measure the physical states of quantum systems. Mathematically, the uncertainty principle can be expressed as $\Delta x \, \Delta p \geq h$, where Δx denotes uncertainty in position, Δp denotes uncertainty in momentum, and h is Planck's constant. Stated thusly, the uncertainty in position (x) times the uncertainty in momentum (p) will be equal to, or greater than the Planck constant (which has a near infinitesimal value at $6.62606896(33) \times 10^{-34}$ J·s).

Classicists such as Erwin Schrodinger and Albert Einstein did not take kindly to Bohr and Heisenberg ridding the world of a physical foundation, and sought long and hard to refute their quantum machinations. A raging debate over the nature of reality ensued, splitting the physicist community down the middle, with the older classicists tending to side with Einstein and Schrodinger, and the younger quantum enthusiasts tending to side with Bohr and Heisenberg. The quantum/classical debate ultimately crystalized as a gotcha-match between Einstein and Bohr. This, due largely to Einstein levelling gedanken after gedanken at Bohr in a desperate, albeit well-thought-out attempt to undercut Copenhagenism and salvage Newtonianism. The quantum/classical debate-turned-academic-death-match proceeded via thought-experiment locking both men in a fierce intellectual volley that lasted until their deaths. Regrettably, the debate wasn't properly settled until 1982, long after the deaths of both. (Einstein died in 1955, Bohr in 1962).

Einstein's major answer against the non-locality and indeterminacy of Copenhagenism was local hidden variables.[67] Einstein's theory however, while popular among classicists of the period was effectively refuted by Alain Aspect in the early 1980's, effectively ending the long debate.[68] Aspect not only refuted Einstein's local hidden variable's postulate (and the associated claim that the Copenhagen interpretation was incomplete), but also experimentally confirmed the core features of Bohr and Heisenberg's fantastical Copenhagen interpretation.[69]

---

[67] In 1935 Einstein, along with colleagues Boris Podolsky and Nathan Rosen posited the existence of local hidden variables as a counter against Copenhagenism.

[68] Aspect, along with partners P. Grangier and G. Roger, writing in *Physical Review Letters* 47 (1981) states: "Our results, in excellent agreement with the quantum mechanical predictions…rule out the whole class of *realistic local theories*." (Emphasis mine).

[69] Throughout this work I refer to the "Bohr/Heisenberg" CI. The ideas I present in this work are based on an ontological interpretation of indeterminism as posited by Heisenberg (with a realist, albeit, non-physical understanding of wave-function), rather than the positivistic interpretation of Bohr, which viewed the wave-function as a mere mathematical convention reflecting mere human knowledge. However, it would be wrong to leave the name of Bohr out of CI presented here insofar as Bohr both defended Heisenberg's uncertainty principle and helped lay the anti-physicalistic foundations of quantum theory itself. My use of, and references to the CI are of a general nature, as I adopt it largely for its ontological claims. This is not uncommon. For, even though the CI is the generally accepted interpretation among working class physicists, there remains no strictly defined, universally accepted version of the CI. Only a number of slightly differing CI's rooted in a few core themes. My use of the term "Copenhagen Interpretation" then, while largely Heisenbergian, is of this latter class, with elements uniquely its own. In the place of Bohrian *anti-realism*, for

Aspect experimentally established the empirical foundations of Copenhagenism, but also confirmed what many quantum theorists had already come to suspect—specifically, that the Copenhagen interpretation was more than a mere interpretation among competing interpretations of quantum theory—the CI being far more robust than any mere translation of facts of data. It thus came to be recognized that the Copenhagen interpretation was an accurate depiction of quantum reality.[70]

This, of course is not to say that competing interpretations do not still abound, but only that the majority of working physicists are Copenhagenists for a reason. And that reason is that the CI is an excellent approximation of reality, *as reality is, in itself*.[71] This is as much a testament unto the scientific sensibilities of Bohr and Heisenberg as it is unto the fact that quantum reality is so fundamentally transparent as to be virtually self-interpreting, providing, in itself, so nude a portrait of nature as to be taken as *reality basic*.

And here is where it gets interesting; for, this 'basic reality' is radically contrary to not only the Newtonian paradigm *but also the atheistic implications thereof*.[72] In fact, it is more than contrary, it is

---

instance, I posit quantum *anti-physicalism*, with a realist ontology for wave-functions—which I believe to be ontologically and substantively spiritual.

[70] Even while recognizing that this reality is always approximate and tentative; in accordance with the tenets of critical realism.

[71] It is said to be unwise to link one's theology too closely with science, given the tentative nature of scientific truth. However, if God is indeed the creator of the natural world then theology *must* speak to scientific truth, for it is after all God's truth; if, but a tentative approximation thereof.

[72] Newtonian physics was given an atheistic interpretation by the Hobbists, the philosophes, and the Laplacians, Newton himself however was no atheist, having been mathematically mentored by the English scholar, mathematician, and Cambridge Platonist Isaac Barrow (1630-1677). Under the tutelage and guidance of Barrow and the Cambridge Platonists (who developed a highly Christianized version of Platonism) Newton dedicated himself to meticulous religious studies. The Cambridge Platonists built their metaphysical system of rational faith from a number of philosophical and theological sources, ranging from Plato, Zeno, and Plotinus (the father of Neo-Platonism), to Origen (the early Church Father) and Erasmus (the Dutch Reformer). At Cambridge, these displaced the Puritan divines, but were themselves later displaced by the virtuosi. The virtuosi, led by Newton's close friend Robert Boyle, fought against the French philosophes of the day, who hijacked Newtonian mechanics to foster their own atheistic agenda. Through it all Newton remained steadfast in his faith. Even so, Newton's theology was not entirely orthodox. For instance, while Newton's *General Scholium* asserts Christ to be "Lord of Lords" Newton's views on Christology and the Trinity were (even by liberal accounts) *heretical*. As a textual critic, Newton came to believe that key passages in Scripture were doctored

antithetical! The ontological table has turned. Only this time, the turn is in favor of Christian theism—and most gloriously so.

With Aspect came the verification of the work of Irish physicist John Bell (1928-1990), the subsequent confirmation of quantum non-locality, and the long awaited validation of the Copenhagen interpretation; along with the repudiation of Einstein's realism and his bid for local hidden variables.[73] The data is now in, and the unlikely (indeed impossible) status of quantum reality is just as Bohr and Heisenberg had imagined.

We now know that there is a reality *beneath* reality. An abstract realm beneath the physical realm. And its hallmarks are quantum *indeterminism,* quantum *holism and* quantum *non-locality.* Together, these seemingly implausible processes produce a distinctively theistic ontology and provide the requisite existential ingredients necessary for a *realist Christian reality.*[74]

Anti-Realism and Spiritual Realism

The version of the Copenhagen interpretation set forth in this work *is not* after the early positivist/instrumentalist epistemology of Bohr[75],

---

by Athanasius (296-373 AD) to make Christ equal with God the Father and the Holy Spirit. Alas, Newton rejected both the divinity of Christ and the doctrine of the Trinity. Even so, Newton would become a Fellow at "Trinity College."

[73] Despite having a strong air of contrivance "*non*-local" hidden variable theories (see David Bohm, *Wholeness and the Implicate Order* [Routledge, 1980]) are still very much on the table of quantum interpretations.

[74] In light of the failure of neo-orthodoxy (and the associated Biblical theology movement) to adequately address the modern liberal theological assertion that divine action is subjective, Langdon B. Gilkey (writing in 1961) made the following plea, "what we desperately needs is a *theological ontology* that will put intelligible and credible meanings into our analogical categories of divine deeds and of divine self-manifestation through events." (Italics mine). Gilkey's concerns here are exactly on point. Today, over a half century later, Gilkey's request has been fulfilled, for, quantum mechanics provides just such an ontology. What Gilkey should have perhaps recognized in his '1961 plea' is that Pollard and Heim had already set forth just such an ontology in the 1950's! See Langdon B. Gilkey, *Cosmology, Ontology, And the Travail of Biblical Language; Journal of Religion* 41, [1961] University of Chicago Press. See also Pollard (1958), and Heim (1953).

[75] Early on, Bohr seems to have subscribed to positivism and instrumentalism, the former of which is anti-realist while the latter is best described as non-realist. The latter Bohr however was no anti-realist, but neither was he a realist, at least not in the naïve classical sense. Rather, it might be said that the elder Bohr set forth his own particular brand of 'quantum realism' rooted in the non-classical framework of his complementarity theory, which utilized complementary modeling of associated albeit antithetical phenomena. Due to the "modeling" intrinsic to his theory, Bohr is often (I believe rightly) labelled as

but rather, after the quantum *ontology* of the latter Heisenberg/Bohr, which held that quantum systems, while objectively existent in between measurements, possess no objective physicality[76] in between quantum measurements—albeit, with an added emphasis on (1) Heisenberg's dualistic vision of a physical classical reality underwritten by a non-physical quantum reality; and (2) Heisenberg's non-physicalistic interpretation of Schrödinger's wave mechanics. This particular denial of the physical is less an assertion of anti-realism and more a statement of "spiritual realism."

The Real vs. the Physical

Historians generally label Bohr and Heisenberg as anti-realists. Regarding the Copenhagen interpretation however, the term quantum "non-physicalist" seems to me to be more accurate, in that the wave-function, despite being an objectively real entity, has no physical properties whatsoever. Though abstract, wave-functions are dynamical entities that evolve over time in intelligible ways which can be represented formally. And herein lies the rub, for, anything that evolves and/or changes in a lawfully and mathematically intelligible manner over time *must be regarded as real*, even though it may *not be physical*. There are thus very good reasons for accepting that the wave-function is in fact a *real*, albeit *non-physical* entity.

What's more, assuming that the wave-function is both real and incorporeal seems the best way to account for the predictive and explanatory success of the Schrödinger wave-equation in deriving its

---

a critical realist. Regardless of the proper philosophical tag for Bohr's latter views, one thing is for sure: quantum phenomena is profoundly mysterious. Observed quantum systems can appear as either waves or particles; depending that is, upon the subjectively chosen experimental set-up of the experimenter. Unobserved quantum systems however, are neither physical waves nor physical particles. Rather, the unobserved quantum system exists as an incorporeal, transpatiotemporal 'probability wave' (or 'probability function'), referring to the statistical probabilities generated by squaring the wave-function, (or amplitudes of those potential existents which may or may not physically actuate upon measurement).

[76] While this position may seem to be anti-realist, the reality of the situation is far more nuanced. For, as I believe the data shows, the wave-function is not anti-realist but rather anti-*physicalist* (or non-*physicalist*). Simply put, the wave-function is an objective reality that is not physical. And this suggests the existence of a substantival essence that is (ontically speaking) either platonic, noumenal, spiritual, or all three. Either way, material monism is dead and a new metaphysically-friendly account of reality must be sought.

statistical solutions. Whatever the substantive status of Ψ may be, there can be no doubt that Ψ accurately represents, and/or very closely approximates an objectively real entity that evolves deterministically over time in a mathematically intelligible manner.

The fact of the matter is simply this: *quantum mechanics works*. And it works with a degree of precision unmatched in the history of science. How is science to account for this great success? The best explanation is simply that the ontological structure of *quantum reality* as presupposed in the dynamical laws of *quantum theory* closely approximates the actual structural ontology *of reality*.

So then, against the classical realism of Einstein/Schrödinger, and against the positivistic anti-realism of say, the early Bohr, many today posit a form of *critical realism*[77], which holds that "epistemology models ontology"[78] such that (contra early Bohr) Psi (Ψ) speaks to an objectively existent wave-function that (contra Einstein/Schrödinger) is non-physical—and hence incorporeal. Quantum physics thus represents a meta-physical sphere that is ontologically fundamental— not to mention ontologically *prior to* the physical reality that sits atop it (while also emerging out from it).

The positivism, and hence anti-realism, of the early Bohr seems to mistakenly mingle epistemology with ontology when it states that only that which is empirically verifiable and/or experimentally demonstrable (or observable) exists. This radical brand of empiricism presupposes, and thus smuggles into science, a material monist ontology that is metaphysically unjustified. Similarly, the classical realism of Einstein and Schrödinger mistakenly conflates what is physical for what is real. Anti-realists on the other hand do just the opposite, assuming that what is real is necessarily physical.

I posit a middle-ground wherein, ontologically speaking, that which is real isn't necessarily physical and that which is non-physical isn't necessarily unreal. Epistemologically speaking, that which can be made known, and thence said to exist, isn't necessarily physical, and that which is non-physical can nevertheless be real and thence made known.

---

[77] For more on critical realism see the associated works of Roy Bhaskar and John Polkinghorne.
[78] The phrase "epistemology models ontology" was famously coined by John Polkinghorne as an expression of his own critical realist philosophy.

Quantum realists and quantum anti-realists will continue to talk past each other until the ontic/epistemic equivocating of both sides is realized, addressed, and resolved.

Bohr and Beyond

Today, many quantum thinkers are setting realism and anti-realism aside for the middling path of "critical realism." Critical realism (and more specifically my own brand of critical *meta*-realism) allows one to recognize the legitimate existence of both corporeal and incorporeal quantum realities, whether these be physical and corpuscular or non-physical and wave-like.

Regarding then, the quantum precept that "quantum systems have no *physical reality* in between measurements", it is quite true, however this does not mean that unmeasured quantum systems *cease to exist* as is often assumed. In reality, (that is, "quantum reality") physical reality can and does give place to abstract/platonic reality, and vice versa. That this latter abstract/platonic reality is in fact *an actual reality* is demonstrable (again, see wave-function interference patterns). Wave-functions, I know, are difficult to fathom. However, what we do know is this: (1) They are mathematical entities. (2) They are governed by dynamical laws. (3) These laws approximate a reality that is structurally wave-like, substantivally abstract, and ontologically platonic.[79] And after this, we know next to nothing. For, here is a reality that is literally beyond us—both tangibly and eternally. And yet, this we do know: whatever this abstract mind-like substance is, it is that which breathes the fabled fires of being into the equations that 'not only' make a universe, but make a distinctly (*and ontologically*) "Christian" universe at that.

My particular claim for a realist (albeit non-physicalistic) wave-function ontology, while not quite the same as the anti-realist accounts of Copenhagenism, does share in the Copenhagenist rejection of wave-function physicalism. In a nutshell I subscribe to a rather straight forward critical-realist-inspired account of the wave-function, which rejects the *physicalistic* realism of the Schrödinger ontology, and in its place adopts the rather obvious alternative; namely: *platonic realism*—replete with the

---

[79] Throughout this work I use the words "abstract", "platonic", and "spiritual" interchangeably in reference to the essence and ontic of wave-function.

abstract ontological status so blatant within the quantum mechanical formalism itself.

Regarding wave-function (Ψ) ontology; whereas classicists posited physical realism, and Copenhagenists posited anti-realism, I posit a form of platonic realism wherein the dynamical properties and structural ontology of Ψ exist objectively, albeit incorporeally as an abstract/noumenal essence located within and occupying a transcendent nomological domain within a space all its own.[80]

As Max Born taught us, wave-functions are dynamic probability fields. However they propagate neither in three-dimensional space, nor in the four-dimensional spacetime array of classical field theory. Rather, multi-component wave-function systems propagate in the vastly higher-dimensional domains of abstract configuration space; *which, space, is itself wave-function,* being an abstract, transcendent, platonic-like realm/medium that is both lawful and mathematical, as well as structurally and dynamically wave-like.

(**Sidebar:** On a different but associated point; some may conclude that there is a hidden suggestion here that spiritual reality is wave-like. I strongly doubt that this is the case. Even so, where spiritual reality overlaps and/or operates at the quantum level it is in fact wave-like. Here however, the wave-like structure is purely functional, operating in the

---

[80] In formalisms, multi-particle or multi-component wave-functions are represented as abstract mathematical constructs propagating in the higher-dimensional-domain of configuration space. The question is: what do these mathematical formalisms refer unto? Classical wave-function realism understands psi (Ψ) as referring to a genuine physical reality in space-time. I on the other hand (following the revolutionary insights of quantum pioneer Max Born) understand psi (Ψ) as referring to an abstract, evolving (i.e., propagating) probability field whose probability amplitudes, while stochastically derived, (and hence nomic in character and structure) are nonetheless objectively real. Quantum wave-functions evolve deterministically but collapse to an indeterministic outcome. It is in the midst of the latter (i.e., collapse) that the mysterious act of a-causal quantum actuation happens. Collapse then, results in the indeterministic a-causal actuation of a quantum event. This self-same instant marks the moment when one of a potential myriad of possible quantum outcomes is physically actualized. These "potentialities" (being mathematically generated solutions to the Schrödinger equation) are approximations to an objectively existent platonic reality. Wave-functions, although mathematically generated in higher-dimensional domains of configuration space, are not simply mathematical "tools", they are objectively existent realities. They are platonic domains, being seemingly an odd mixture of mental construct, spiritual essence, and theological necessity. And much more, for, they are the underwriting nomological constructs necessary to a theological reality.

capacity of a structurally dynamic formula for generating probabilistic existents (which of course may or may not be actualized upon wave-function collapse).

Bohr/Heisenberg Copenhagenism, while denying the existence of an unobserved/unmeasured *physical* quantum reality, *does not* necessarily deny the existence of an unobserved/unmeasured *abstract* quantum reality—such as that mysterious waving reality/entity unto which the solutions of Schrödinger's wave-equation so accurately refer. The question, ultimately, is what status are we to ascribe to the wave-function?

The popular position that wave-functions are mere mathematical tools that help generate probabilities, however, falls way short in light of double-slit interference patterns—which, patterns only derive of actual waves. And since the "interference patterns" are actual, so must be the waves that produce them. It is not a difficult leap to reason from wave-functions as tools to wave-functions as realities. In fact, in light of the empirical evidence, this leap is effectively required.

Even so, wave-function realism (of the platonic sort) remains a tough sell in the material monist climate of current science. The reason for the obstinance is obvious: once empirical science acknowledges the objective reality of abstract, metaphysical, and/or spiritualistic entities, the game is over. Naturalism has lost. You see, the great boast (and hope) of secular science is to be able to explain everything in purely naturalistic terms. And this necessitates a material monist backdrop. However, if "other" realities genuinely exist; that is, realities that are not "material", that are not "natural", then the game is lost.

Understand, quantum reality destroys naturalism and/or materialism, but leaves science and scientific law untouched! This is possible because naturalism and materialism were never a legitimate part of orthodox science. Rather, these were imposed upon science by atheistic ideology.

Quantum indeterminism for instance, is well within the confines of scientific law, and so poses no threat to scientific orthodoxy. And in this we see a reflection of the brilliance of the Creator who engineered the world in such a way that He can act freely at the quantum level while the macro-world is none the wiser.

The new facts are in: reality is dualistic! For, wave-functions are non-physical yet objectively existent. If they weren't, then the mathematically generated values physicists assign to wave-function amplitudes simply wouldn't work; for they would not derive of an actual reality. For the quantum formalisms to work they must correspond to an actual reality. This actuality is wave-function, and its wave-like structure has been indirectly (but empirically) observed innumerable times in various double-slit experiments, where overlapping wave-functions produce interference patterns—which, patterns are only associated with wave phenomenon.

Moreover, our mathematically derived (and stochastically assigned) values do work, (and work incredibly well) allowing for accurate predications to be made. And this tells us something significant. Namely, that the mathematical models utilized closely reflect the given realities unto which they refer. The evidence, being, the success our models!

Formally, wave-functions are abstract mathematical realities; models, if you will. They nonetheless reflect actual realities. Objective realities that are just as abstract and dynamic as the mathematical models. Here is where mathematically inspired constructs of the human mind, and objectively existent constructs of the Divine Mind brush up against one another. Here is where the platonic and the mentalistic meet. Wave-functions then, are mathematically generated constructs, and objectively existent realities, grounded within a platonic ontic—the former being approximations of the latter.

Incredibly, wave-functions blur the line between abstraction and reality. Not surprisingly then, the language utilized in the mathematical formalisms of quantum theory is used interchangeably with the very reality those formalisms are meant to approximate. The term "wave-function" for instance, can refer to either the formalism, or the reality unto which that formalism refers. This conflation of quantum formalism and quantum reality is, I believe, an unavoidable consequence of the platonic nature of quantum reality.

For such reasons, (and others), we should take quantum reality seriously, and allow *it,* to inform *us,* as to how *it,* actually is. When we do this, that is, take the evidence as we most naturally find it, the evidence reveals that beneath it all, that is, beneath the whole of physical reality, there lies a spiritual reality that is not only foundational to the physical realm, but fundamental to it as well, being the very source from whence 'the physical' existentially derives.

Spiritual Realism

The abstract reality that exists in-between quantum measurements smacks of a metaphysical or spiritual reality. This foundational reality in turn underwrites *all* physical reality. I thus posit a reinterpretation of the Copenhagen interpretation exchanging anti-realism with what I call "spiritual realism." I contend that the quantum reality that exists in-between quantum measurements, and which ultimately gives rise to physical reality itself, is no less than the metaphysical reality from which the fundamental spiritual essences of soul and spirit (and perhaps in some way 'mind') ultimately derive—albeit, by and through the creative will of God.

This is of course, speculation, but it is informed speculation that is suggested in, by, and through the world's ontological character and structure. (Again, recall wave-function interference patterns). As for soul/body interfaction, recall the structure of the brains synaptic junctures, the associated quantum processes, and the inherent implications for mind-brain/soul-body interfaction.[81]

All in all, my assertions of quantum reality are not altogether that dissimilar from anti-realist assertions, the main distinction being my understanding that quantum reality is more anti-physicalist than anti-realist, insofar as an abstract reality can indeed be real, such as in a platonic sense. Again, the error of anti-realism, being, that it equates "realness" with "physicalness", in turn leaving no ontological space for *non-physicalness,* which is exactly what quantum reality demonstrates (and Christianity necessitates).

Early quantum theorists such as Bohr and Heisenberg correctly identified quantum reality (i.e., Ψ) as being abstract, but failed to recognize this abstract reality as constituting an objectively existent

---

[81] Quantum-based consciousness was explored early-on by thinkers such as Karl Heim, Arthur Compton, et al., and continues to be explored today (see the work of Henry Stapp, Penrose and Hameroff, et al.).

reality, realm, or substrate.[82] Even so, one can only imagine how odd it must have been to hear Bohr speak on the issue for the first time.

For instance, in the now famous Como address of 1927, Bohr stated his conviction that the entities of quantum-level reality (i.e., waves and particles) are but "abstractions, their properties being definable and observable only through their interaction with other systems" (e.g., classical-sized measuring instruments). The implications of Bohr's address must have been mind-blowing (especially to those classicists in the audience).

What Bohr was suggesting after all was that sub-atomic entities have no physical properties until those (hitherto non-existent) properties are properly measured! The implication being that prior to measurement these properties (and hence the entities themselves) exist only as abstractions—yet are somehow made real (i.e., physical) via classical interaction! Bohr's assertions however, only makes sense if the abstract quantum reality has an objective existence, such as a platonic existence. Otherwise, objective reality derives purely and entirely from subjective realities—namely, those conceived mathematically in the minds of quantum theorists! This however, is patently absurd, quantum reality therefore must literally exist!

In closing then, let me again reassure the reader that, despite the ambiguous nature of the quantum ontic[83], the existence of an abstract, mind-like enspacement existing just beneath material reality is an empirical scientific fact!

This being the case, die-hard Newtonians need to cease from attempting to explain away the implications of quantum reality (*via* ideologically-inspired claims of incompleteness rooted in classical

---

[82] The real but non-physical nature of wave-function has led to much confusion. Heisenberg himself spoke of quantum reality in seemingly both anti-realist and realist terms (see for instance Heisenberg's 1927 paper *Quantum Kinematics*). This shifting back and forth between anti-realism and realism is understandable considering that the wave-function is simultaneously realist and anti-realist—the caveat being that the term "anti-realism" has mistakenly been applied when in reality we are dealing with an anti-physicalism!

[83] Quantum reality/ontology is enigmatic to be sure, however, there is, I believe, a very good reason why quantum reality/ontology is as mysterious as it is. And it has everything to do with the Christian understanding of the world *as creation*. You see, the Christian conception of creation is a Trinitarian conception, with Christ the *Logos* recognized as the chief creative source and rationale behind all things. The world is Christ's plan, and so through Christ comes the world's redemption. Christ then, is both Creator and Redeemer of this world. Little wonder, "volition" (the very basis of "faith" and "free will") turns out to be ontically basic to this reality.

nostalgia and a longing for a return to deterministic science) and start believing what quantum physics straightforwardly tells us: that there exists a platonic under-reality from whence the whole of physical reality emanates.[84]

---

[84] Heisenberg gave the world (1) *quantum indeterminacy*, (2) *wave-particle dualism*, and (3) *abstract wave-function realism* (i.e., the concept that wave-functions are non-physical but ontologically real). The metaphysical implications of these three quantum precepts are as incredible as they are theistically-charged. Together, these speak to a reality that is uniquely Christian in character, being ontically tripartite in both structure and relationality. Within Heisenberg's work can be found the elements of a robust theological metaphysic that was unknown to Heisenberg himself. This metaphysic is strikingly platonic, particularly with respect to Plato's allegory of the cave (as told in the Seventh book of *The Republic* (366 BC) which speaks of reality as having a tri-level structure. In Plato's hierarchy, God (the divine) is the highest and truest source of being and reality. And from God comes the second source of being and reality: the noumenal realm of cognate intelligibles (wherein exist abstract eternals, mathematical truths, [and, I might add 'wave-functions'] etc.). From this second source derives the third, final, and lowest of the realities: that of the phenomenal—the physical. So then, from God comes the noumenal realm, and from the noumenal realm comes the physical realm. Anyone vaguely familiar with quantum mechanics will immediately see the parallel. Impressively, Plato accurately depicts 21st century ontology from the fourth century BC! We actually live in a reality where the physical derives of the noumenal! And the noumenal from God! (The latter, I assert on the grounds that the noumenal, being ontically prior to that which is natural, must owe to that which is both supra-natural and supra-noumenal. And this can only be God). We live in a universe with a tripartite ontology (perhaps not surprising given the Trinitarian nature of the Creator). This is reflected in both reality and in us. For we are body, soul, and spirit. One may think of it this way: we dwell in the physical, yet have access to the divine (in Christ) through the noumenal (via prayer). (This tripartite nature of reality has not been entirely lost on the scientific elite. See particularly the "Three-worlds" model of Roger Penrose (*Shadows of the Mind*, 1994), and the "Three-worlds" model of Karl Popper (Tanner Lecture, University of Michigan, 4/7/1978).

Chapter Three

Interlude: The Intelligible Cosmos

> The miracle of the appropriateness of the language of mathematics for the formulation of the laws of physics is a wonderful gift which we neither understand nor deserve.
>
> —Eugene Wigner

The Intelligible Cosmos

Quantum reality is beguiling. Quantum reality is double-dealing. Quantum reality is weird, uncertain, counterintuitive, mysterious, and paradoxical. Yet, quantum reality is all these things for a reason. Quantum "uncertainty", for instance, allows for a world where spiritual phenomena can not only *take place,* but *"take place" without infringing upon the natural regularities requisite to science.*[85]

From a "cosmic design" perspective, quantum weirdness is a highly economic means to an end desired by God. This "end" being spiritual phenomenon that doesn't infringe upon physical process, yet which can interface with physical process. When understood this way, quantum peculiarity is an onto-existential cost that is next to nothing.

---

[85] Science is important in that its focus is the "Book of Nature", which book, is a revelation in itself. Recall that Christian tradition has long held that God wrote two Books, the Book of Scripture and the Book of Nature.

Quantum peculiarity, as counterintuitive as it may be, is in all likelihood the most economic, and most rational way possible to realize a Christian realist reality of the sort desired by God. Quantum reality, while speaking to the general creative wisdom of God, also speaks to the particular creative wisdom of Christ in preparing a cosmos with the requisite arrangement of ontological traits necessary to handle the theological demands of the divine plan *in Christ*. You see, Christianity demands a very specific ontology, and quantum mechanics realizes that ontology with elegance and ingenuity.

The various laws, processes, and phenomena of quantum mechanics seem to share a single-minded goal—that of realizing the particular set of theological tenets associated with Christ as both Creator and Redeemer of the world. Christ, however, is also the eternal *Logos*, the Author of creations rational intelligibility. And the success of the scientific enterprise is a testament unto this rational intelligibility. Particularly as witnessed in the stunning, and seemingly inexplicable degree of correspondence that exists between the intrinsic rationality of the subjective human mind (which so efficiently models objective reality) and the intrinsic rationality of the objective natural world (which, in turn, reflects the rationality of the Mind of God).

The rationality of the Mind of God is reflected in the rationality of the created order, and is only intelligible to the mind of man because only man is created in the image of God. Science then, rather than being an atheistic endeavor (as many materialists proclaim and many fundamentalists parrot) is an occasion to fathom the divine Mind. Johannes Kepler expressed this very truth when he described his scientific efforts as "thinking God's thoughts after him."

That the human mind is so adept at penetrating the complexities of nature suggests a degree of correspondence between the rationality of the human mind and the intelligibility of the cosmos (which, intelligibility, is a reflection of the rationality of the divine creative Mind). While remarkable, this seeming one-to-one correspondence between the rationality of the human mind and the intelligibility of the cosmos is perhaps not surprising, considering that the divine Mind is the Author behind both.

There is an inexplicable consonance between the world of abstract human thought and the world of concrete cosmic structure. And the success of the scientific enterprise is a testament unto this fact. The intellectual power and prowess of the human mind is far beyond anything that would have arose via evolutionary selection pressures. There simply were no selection pressures on early humans to produce brains capable of manipulating highly abstract concepts such as are necessary for doing higher mathematics or understanding quantum physics.

Natural selection after all, is purely utilitarian; that is, it sculpts species for utility—for raw survival and reproduction. Higher human endeavors such as poetry, art, literature, music, or mathematics are simply beyond its scope. Nature simply has no tools to sculpt minds towards these ends. There simply are no selection pressures to produce a Bach, a Newton, or a Michelangelo. There are no crucibles of nature to forge a Shakespeare, a Einstein, or a Leonardo.

From an evolutionary standpoint there is simply no reason why the human mind would need to possess the capacity to either invent calculus, solve the Schrodinger equation[86], or be able to calculate Pi to the trillionth decimal place! Evolution adapts to immediate environment. Why then, is the human mind so well adapted to the exceedingly remote, incredibly abstract spaces of higher dimensional mathematics and physics?[87]

When confronted with this mystery, evolutionary materialists unimaginatively reply that these higher mental capabilities of man were simply a fluke. This however solves nothing. After all, what good is the ability to do higher mathematics *prior to the invention of mathematics!*

Furthermore, however imaginative (or unimaginative) such explanations might be, they do nothing to explain the mysterious one-to-one correspondence between the rationality of the human mind and

---

[86] That the human mind is able to fathom the abstract sphere of quantum reality is every bit a mysterious as quantum reality itself. Think about it; why should Darwinian-evolved humans, sculpted exclusively for carnal survival, be mentally equipped to navigate the radically counter-intuitive contours of the quantum landscape? Especially when there is nothing even remotely analogous to that landscape in nature! Because our minds evolved at the classical level, there is nothing in nature that could have prepared or equipped us for the challenges of quantum reality. But this is just part of the mystery. When it comes to the human mind, it is not simply that the whole is beyond the sum of its evolutionary history, it is that the whole is so distinctly (and suspiciously) fitted for grasping at meta-realities! How is it that we are so sensitive to the transcendent—the eternal—the divine? From whence comes this eternal yearning after God?

[87] I am not here suggesting that our brains did not evolve. Only that Darwinian evolution (of the brand offered by materialists) is insufficient to account for key aspects of the human mind.

the intelligibility of the cosmos. Why the uncanny intellectual fitness of the former to the subtle ingenuity and nuanced complexity of the latter? It is as if the mind was intentionally appropriated to plumb the abstract depths and discover the profound truths of the cosmos! But why? Why this intimate correspondence between the degree of transparency of the cosmos and the comprehension level of man? This is a question to which Darwinian evolution simply has no good answer. The reason being, I suspect, is that it has less to do with selection pressure and more to do with the doctrine of the *Imago Dei*.

Christian tradition has long held that nature is a "Book" to be read. If this is true, that is, if nature is indeed a revelation of God, then man is its sole interpreter! *Could it be that quantum reality, insofar as it panders to Christian doctrine, is revelating unto man that the true nature of nature is that of divine creation?*

I am persuaded as much.

Eugene Wigner famously spoke of the "unreasonable effectiveness of mathematics"[88] for describing the physics of reality. This is indeed a mystery. One magnified by the fact that nature's structural arrangements are best reflected by those mathematical equations and formulae that are the most beautiful, the most elegant, the most economical, etc.

This mystery is profoundly suggestive. For, it suggests that there is a common creative Mind behind (1) the world, (2) the mathematics of the world, and (3) the mind of man—which can not only fathom (1) *and* (2), *but also the existence of the creative Mind behind all three!*

No worldview accounts for the inexplicable correspondence between human rationality and the rational transparency and beauty of the cosmos better than Trinitarian theism, wherein Christ the eternal *Logos* creates man in his image to dwell in a cosmic tabernacle that is nothing less than a revelation of the divine Mind itself.

Little surprise then, that quantum reality, which is the ground of physical reality, leads straight to abstract rationality that is itself a kind of logos. Just not "*the* Logos", or 'personified rationality' from whence comes cosmic intelligibility and hence scientific rationality to begin with.

---

[88] See, for instance, Eugene Wigner's 1960 paper entitled "*The unreasonable effectiveness of mathematics in the natural sciences.*"

The Quantum-based Christian realism presented in this work then, provides no-less than a self-contained, internally-consistent worldview that both begins with Christ as Creator, and continues eternally in Christ as redeemer in the eschaton!

Chapter Four

The Medium of Divine Mind

> We believe in one God,
> the Father, the Almighty,
> maker of heaven and earth,
> of all that is, *seen and unseen*. (Italics mine)
>
> —Nicene Creed

Quantum Reality

Let us now pause for a moment and take stock of what we have thus far learned.

As we have seen, quantum wave-functions collapse to give rise to physical quantum particles having specific, well-defined values/states/properties, etc. These particulate states however do not persist indefinitely. When quantum particles cease from interacting with classical reality they in turn cease from corporeal existence returning instead to their incorporeal wave state.[89] Isolated quantum particles

---

[89] Quantum event outcomes actualize upon wave-function collapse. Wave-function collapse, in turn, occurs when a wave-function interacts 'irreversibly' with a macroscopic object. These outcomes however, while necessarily caused, are insufficiently determined. Quantum events are therefore said to have necessary but insufficient causes. God then, does not act as "causer" of quantum events, for, such events have necessary natural/mundane causes. Rather, God acts as "determiner" of quantum event *outcomes*, "willing" and/or "determining" through special providence what are otherwise underdetermined quantum events. This is not to say however, that God determines all quantum event outcomes.

instantly reassume their abstract wave-function status, and continue so, until once again coming into contact with a physical macroscopic system. Between interactions (i.e., measurements) wave-functions simply evolve (i.e, propagate outwardly), and do so continuously and deterministically in accordance with the governing dictates of the Schrödinger wave-equation.

Mysteriously, wave-functions, being deterministically governed, evolve (i.e., propagate) both temporally and spatially; albeit, not in the spatiotemporal dimensions of this physical universe. Rather, they do so within the eternalistic dimensions of the platonic universe—the abstract ontology of which, provides an objective reflection of the metaphysical status of quantum reality. These abstract waves, as objectively modeled, ultimately serve as both the means and mechanism of physical reality, not to mention the various doctrines of the Christian faith.

Mathematically, wave-functions are designated by the Greek letter psi (Ψ). Psi squared, (or Ψ^2) represents the square of the wave-function amplitude and gives a probability density for the complementary quantum particle physically actualizing at various locations upon wave-function collapse.[90] (Recall that wave-functions evolve deterministically, but the outcome they collapse unto is indeterministic—i.e., apart from probabilities generated during wave-function propagation—at which time the statistics evolve deterministically).

Quantum entities thus have a physical side and an abstract mathematical side, with the latter being every bit as real as the former; indeed perhaps even *more real,* as platonic tradition suspects, and as Christian theology suggests. As regards the latter; Christ is the eternal *Logos,* or pervading rationality behind and beneath the intelligibility of

---

But only that the open ontological structure of quantum process is uniquely suited to allow a disembodied being such as God to determine such outcomes if He so desired. I personally believe that God acts to determine only some quantum event outcomes. To say that God acts to determine all, would be to suggest a form of divine determinism; thus raising issues of theodicy. On the other hand, to state that God determines only "some" quantum event outcomes begs the question: what determines those underdetermined quantum event outcomes that God does not act to determine? I would suggest that God may still determine these, albeit non-specifically. God may allow these to be randomly determined in accordance with their stochastic limitations. This would be a case of God's general providence, with God acting to sustain the stochastic processes that underwrite quantum reality. Conversely, it is a case of special providence if God acts with specificity to determine a particular quantum event outcome.

[90] Quantum wave-functions, though themselves spatially transcendent (being abstract platonic entities) nonetheless specify (albeit stochastically) for real-world states/locales.

the cosmos. Nowhere is this rationality more clearly manifest than in the underwriting mathematics of the cosmos. Particularly as reflected in the quantum wave-function, which, as both an objectively existent entity, as well as a purely abstract mathematical construct, gives rise to the physical constituents that make-up the material world.

The picture of reality that emerges is that of a material universe emanating into physical existence fundamental-particle by fundamental-particle, via the abstract structure and mathematical intelligibility of quantum wave-function. This abstract mathematical substance is the fundamental "stuff" from which the concrete universe was created and continues to be created. Mysteriously, (and indeed, suggestively) this elemental quantum substance from which all physicality derives is far more 'mind-like' than it is matter-like. A fact that has led serious scientific thinkers (from von Neumann and Wigner in the 50's to Chalmers in the present) to suspect that it may play a fundamental/constitutive role in consciousness itself. Not an unwise scientific estimation considering the seeming imperviability of mind and consciousness to material reduction.

The mysterious quantum substance then, while serving as both the means and mechanism of God's general providential sustaining of the created order, may also serve, in one way or another as a missing medium linking mind and matter.

Abstract Wave-function Realism

As we have seen, the corresponding complementary[91] side of a physical quantum system is its abstract wave-function, which is a mathematical function that has the structure and properties of a wave (albeit these properties [such as amplitude] have mathematical and/or probabilistic relevance).

And as we have also seen, quantum wave-functions possess an objective reality; one that is as abstract as the mathematical constructs themselves, yet altogether realist! Here then, is an objectively existent

---

[91] Whereas Bohr tended to focus upon how quantum waves and quantum particles "complement" one another in quantum theory, (suggesting no ontological conflict whatsoever) Heisenberg tended to focus on how waves and particles actually "conflict" with each other ontologically. (cf., Kristian Camilleri, *Heisenberg and the Interpretation of Quantum Mechanics* (Cambridge; Cambridge University Press; 2009).

mathematical construct somehow grounded in a mentalistic and/or conceptualistic reality!

But what can this possibly mean?

Why should something so strange, be so?

And what, if anything, is this telling us about the ultimate nature of reality?

The answers to these questions lay in wave-function reality itself. So let's begin there.

Regarding "wave-function reality", I am an 'abstract wave-function realist'[92] in that I believe the evidence unambiguously demonstrates that quantum wave-functions, while incorporeal, are nonetheless objectively existent, *realist* entities. Wave-functions are composed of a purely abstract substrate. We know this because when two independent wave-functions coincide so as to superimpose themselves one upon the other, constructive and deconstructive interference occurs between their respective crests and troughs. And incredibly, this overlap results in detectable interference patterns that leave physically detectable signatures under laboratory conditions![93]

Wave-functions then, though substantively incorporeal, are capable of interacting with themselves in an objective, albeit non-physical manner. Here are abstract entities interacting with, and interfering with one another in predictable, repeatable way!—thus demanding some or other form of platonic realism.

---

[92] By this, I mean 'Copenhagen realist'; or one who believes that quantum indeterminism is ontological and that wave-functions are objectively (albeit non-physically) real entities. My interpretation of quantum reality, while theologically tinged, is basically Copenhagenist (of which, there are numerous versions). As Max Jammer recognized "the Copenhagen interpretation is not a single, clear-cut, unambiguously defined set of ideas but rather a common denominator for a variety of related viewpoints. Nor is it necessarily linked with a specific philosophical or ideological position." (Jammer, *Philosophy of Quantum Mechanics* 1974). So then, while the theological suppositions of my interpretation are obviously not standard CI, the foundations upon which these stand are.

[93] See also, those inexplicable double-slit interference patterns produced by and through the slow build-up of individual particles fired in slow succession.

Understand just how radical this discovery is. Phenomenologically speaking, prior to measurement, colliding quantum wave-fronts behave just like colliding physical wave-fronts, except that the crests and troughs of the former are no more tangible than even the most vivid figments of a dream.

Quantum physics has revealed that we live in a world with two distinct; two objectively existent realms of being. One physical; the other metaphysical! And this is *not* opinion. This is *not* interpretation (scientific or otherwise). This is objective, empirical, hardcore scientific, 21st century *fact!* Something "other" than just the material reality exists. We may not know exactly what this "other" is (and may never know), but the fact remains: "the whole" is more than just material reality. For, material reality, as we now know, emerges from "mind-stuff" *and not the other way around!*

Quantum dualism is as *profitable* to believers as it is *ruinous* to material monists. Both for its ontological insight, and for its striking fitness as a facilitator of Christian doctrine (or should I say, objective theological phenomena). The question is not simply "why should an abstract essence objectively exist?" As puzzling as that question may be, but rather, "why should an abstract essence objectively exist *that panders so particularly (and thoroughly) to the spiritual demands of Christian theology?*"

Is this not incredibly suggestive?

Why is cosmic law so precisely fitted to a particular theological landscape?

Are we, here, dealing with some form of metaphysical, spiritual, (or dare I say, "Christian") fine-tuning?—*as appears to be the case!*

Could it be that the cosmos has been "frontward-engineered" in anticipation of the metaphysical and ontological demands of Christian theology?—*As appears to be the case!* Is it possible that the underwriting quantum substance is in reality a metaphysically-meted foundation set by God to serve as an ontological ground for the divine theological will?—*As appears to be the case!*

To each of these, the answer is a resounding yes! Here are two instances where the appearance *is* the reality! Quantum reality *appears* to serve metaphysical ends because quantum reality *does* serve metaphysical ends. Simply put, quantum mechanics is metaphysically inclined. And the inclination is toward theological realism. For, the very structure of quantum reality is ideal as a construct for theological realism.

Let me pause here to state that these are extraordinary claims, I know. Yet the fact of their reality is more than attested unto in the fact that the quantum revolution and the (theistically-inclined) tenets thereof, literally reversed each of the (*atheistically-inclined*) tenets of modern (Enlightenment-inspired) science—or should I say "scientism."

Enlightenment science told us that reality is monistic (when it is in fact *dualistic*); it told us that reality is deterministic (when it is in fact *indeterministic*); it told us that reality is reductionistic (when it is in fact *holistic*); it told us that reality is *localistic* (when it is in fact *non-localistic*)! The quantum revolution was more than just a changing of the paradigmatic guard of science: it was the victory of "theistic science" over the three-hundred-year reign of atheistic scientism!

Understand, it is not simply that each of the atheistic tenets of Enlightenment science was overturned by the new quantum ontology; it is that each such tenet was so idyllically replaced with an 'ontologically antithetical' *theistic* tenet!

The Two Realms

Quantum reality exists beneath classical reality. That is, beneath the elusive quantum/classical border where classical (macro) reality ends and quantum (micro) reality begins. This elusive border, which in reality is the boundary between physical reality and abstract reality, has so far eluded physicists.[94] Boundary whereabouts notwithstanding, what we appear to be dealing with here seems to be no less than the dividing line where physical reality ends and metaphysical reality begins; the latter being the locus from whence the divine Mind acts (via the intermediary

---

[94] The real question is how are we to decide where micro-reality ends and where macro-reality begins? Experimental physics has made this question even murkier. For instance, quantum superposition has now been observed in objects large enough to be seen virtually with the naked eye (e.g., the aluminum nitride mechanical resonator measuring approximately 40 μm in length and composed of nearly one trillion atoms).

substrate of wave-function) to uphold the universe in its continued existence.

The discovery of quantum reality has drastically altered the scientific, philosophical and metaphysical landscape. The Judeo-Christian notion of a Creator God that upholds and sustains the cosmos can no longer be sloughed-off as pre-scientific religious nonsense. Quite the contrary, as we enter the 21st century we do so with empirical evidence that the material cosmos emerges from and indeed persists upon an incorporeal essence with an inimitable semblance to consciousness!

What greater scientific confirmation of the existence of the Judeo-Christian God could a believer possibly ask for? I can't think of one.

Mind, Mathematics, and Matter

According to the famous mathematician and infamous atheist Bertrand Russell "pure mathematics brought him nearest to a truly mystical experience of a pure world, deathless and beyond the grasp of time."[95] The quantum wave-function is our link to this pure world. And in like turn, it is God's link to initiating physical causation in nature. But this isn't all. For, as we have seen, it is also the means and mechanism by which God maintains the continuous on-going act of creation.

As we have seen, the physical matter and energy constituents of the universe are continuously brought into physical being via a non-physical mind-*like*[96] essence: wave-function; which, as an abstract entity, generates something akin to active information (in the form of evolving potentialities) manifest as mathematically governed wave-like structures.[97]

---

[95] Bertrand Russell, in Ward, *Pascal's Fire* (London: One world Publications, 2006).
[96] The emphasis here being on "like", because wave-function is not "mind" itself. However it is *mind-like* enough to serve as an intermediary between the divine Mind and creation.
[97] These deterministically evolving potentialities constitute the abstract structure of quantum wave-function. Upon collapse, these statistical probabilities (which recall mark the likelihoods of potentiality actuation) seem to instantaneously condense into a physical manifestation; particulate matter. This uncanny Aristotelian shift from potentiality to actuality seems to suggest that physical existents (such as quantum particles) comes into being via the condensation of "potential" existents. Could it be that the particulate existents of physical reality condense into being, from the distilled sum of divine middle-knowledge for that particular existent (or what is the "wave-function-equivalent" of divine middle-knowledge for this reality).

This is simply odd. It is odd that an abstract entity should have structure—intelligible structure; dynamic structure; rational, predictable structure! Yet this is what is. This is the reality; the fact of the matter. Our universe persists through the abstract dynamisms of a mathematical sea. The universe is literally derivative of mathematics, that is, "physically" derivative of mathematics! But it is not just the symbols and digits that we are talking about. It is a *mathematical reality*, with structure, rationally intelligible structure that is objective and whose behavior is predictable.[98]

So while the physical cosmos emerges and/or emanates from a dynamic sea of objectively existent mathematical structure; this mathematical sea itself, (so it would stand to reason) emerges and/or emanates from "mind" (or rather "Mind"). Mathematical truths may be eternal, but active mathematicisms are conceptual. And mathematical conception requires consciousness, or mind (or "Mind").

What's more, we must remember that mathematical equations by themselves *cannot* generate a cosmos![99] Moreover, those mathematical equations which seemingly do generate a cosmos (such as those evolving and collapsing at the quantum level) *cannot* generate themselves! What appears to be taking place then, is that God, the divine Mind, is cogitating the cosmos into existence, wave-function-by-wave-function.

God therefore, is deriving the underwriting equations of the cosmos, as well as breathing the requisite divine fire into them, providing a universe for them to describe!

A hierarchy of both being and dependence suggests itself. There exists a physical world (the classical world we experience as ordinary, everyday reality). This physical world is ontologically dependent upon a non-physical world that exists beneath it—the quantum world. And beneath, or behind this quantum world there exists the ultimate reality: God—the Divine Mind! Is it any wonder then, that science has discovered that the ultimate foundation of physical reality is "abstraction?" That is, after all, what *minds* do; they *abstract!* Only in

---

[98] Up to the point of wave-function collapse of course. Remember, wave-function evolution is deterministic, but wave-function collapse is indeterministic.

[99] A sentiment I (and countless others) share with Hawking, who, musing on the mathematics of a unified theory asked, "What is it that breathes fire into the equations and makes a universe for them to describe?" (Hawking, *A Brief History of Time*, 1988).

the case of the Divine Mind, abstractness and concreteness are but a mere act of Will apart.

Pause for a moment to consider what is actually taking place. Beneath the physicality of the material world, countless wave-functions evolve, propagating, generating abstract potentialities; possibilities for actuation in the physical reality above. A "physical reality" that is quite literally being 'imagined' into existence, moment-by-moment, from beneath a purely conceptual sea of undulating possibilities.[100]

Reality is ultimately an ontological hierarchy that is wholly and entirely dependent upon God. In this hierarchy we glimpse both the structure of reality and the very construct of creations blueprint. It is almost as if God is 'dreaming' the cosmos into existence, cogitating its material constituents into being particle-by-particle. Whatever the reality, this image is not far from it, insofar as quantum reality is quite conceivably a series of divine abstractions—yea, abstractions that beget concretions! Concretions which collectively comprise our physical universe.

All this is tremendous news for theists. What's more it is news that is long overdue. For almost three centuries the Christian faith has weathered a seeming non-stop barrage of secular scientific challenges. And atheistic materialists have been there every step of the way seeking to undermine the faith and exploit its every error. Now however, the tables have turned. Now, it is the materialists who "need to face scientific reality." Now, it is the materialists who "need to be intellectually honest with themselves and follow the scientific facts wherever they lead, no matter how uncomfortable." Now, it is the materialists who "need to put scientific truth first and personal ideology last." Now it is the materialists who need to "put away childish things and see reality as it genuinely is"—*instead of merely how they wish it to be! Which is deterministic, reductionistic, localistic and monistic!*

---

[100] So, while I am not saying that the wave-functions that make up quantum reality are themselves *the* divine Mind, they very well might be abstractions *of* the divine Mind. This might explain why wave-functions reflect such distinct semblances of divinity; such as the appearance of omniscience, omnipresence and middle-knowledge (via quantum entanglement, non-locality and other quantum phenomena). Other similarities have been pointed out as well. See for instance Polkinghorne, *Quantum Physics and Theology: And Unexpected Kinship* (Yale University Press, 2007) and Ernest L. Simmons, *The Entangled Trinity, Quantum Physics and Theology* (Minneapolis: Fortress Press, 2014).

The reign of secular materialist science is over, whether secular materialist scientists are willing to acknowledge it or not. Cosmic reality, as science presently reveals it, is not only friendly to Christianity, *it is fitted for Christianity!*

One can only imagine how scandalous (if not traumatic) this must be for atheistic materialists after having had things their way for so long. Indeed, modern science has pandered to atheistic materialism for so long that materialist scientists left off long ago from even considering that things might one-day show to be different. But this is exactly what happened when the quantum revolution took place. However, things didn't just show to be 'different' from atheistic science; things showed to be *antithetical* to atheistic science! And the means by which all this occurred was the discovery of the quantum ontology.

The heart of the quantum ontology is quantum wave-function; which I described above as being akin to divine abstraction. Understand that even after a century, science still does not know what wave-function actually is. Nor can they ever know. For, the moment it is gazed upon it is gone. Even so, wave-functions possess unmistakably noumenalistic properties. Even such as to suggest the presence of an underwriting mentalistic source, much the same way that a dream sequence suggests the presence of an underwriting mind.

However, whereas the latter may also suggest the presence of an underwriting brain, the former does not! Recall that quantum theory has taken the materialist presumption that mind comes from matter and completely turned it on its head. For it is now an empirical fact that all matter and all physicality comes from that which is incorporeal, noumenal, and perhaps even "spiritual."

Wave-function Causation

As we have seen quantum wave-function is a metaphysical substance unlike anything we are aware of. Wave-functions, and/or the wave-function substrate, is metaphysical. This much is fact. The question is whether it is more like a spiritual substance or a mind-like substance? Either way, wave-functions exists objectively. Again, this we know empirically. For, wave-functions can be causally efficacious in experimentally demonstrable ways. And if science tells us anything it is that that which has objective causal power is objectively real.

A mentioned earlier, wave-function efficacy is best demonstrated in double-slit interference patterns. Here, intersecting wave-functions superimpositionally interfere with themselves resulting in both constructive and deconstructive interference. This 'interference via overlap' generates a distinctive pattern which can be scientifically observed. Here then, is a transcendent, metaphysical essence capable of producing (and more importantly *reproducing*) a physical signature (i.e., alternating bands of light and shadow) under controlled laboratory settings.

In double-slit interference patterns we have empirical scientific proof for abstract wave-function causality. And in this, we have empirical scientific proof for metaphysical causation! Again, whether this metaphysicality is spiritual or noumenal is impossible to tell, but I have a suspicion that it is both. It appears that the march of science has unwittingly strode across the path to spiritual and noumenal causation!

Whether quantum physicists realize it or not, they are bumping-up against an eternal realm.[101] A transcendent realm long intuited to exist by mankind. Mankind seems to possess an implicit knowledge of divinity—of transcendence—of things spiritual—of things 'beyond.' This knowledge appears to be instinctual within us, existing across diverse cultural boundaries, and from ancient times to the present.

Transcendent reality, which has long been known through intuition, has now been found existing both beneath and behind the material veil of this physical reality. This is not to say that physical reality is simply a mirage of appearance in an idealist/solipsist sense; rather it is to say that physical reality, far from being the whole of reality, and far from being the "stuff" from which consciousness is made, is but one aspect of a vaster meta-reality—where consciousness gives rise to physicality *and not vice versa!*

The discovery of quantum reality is unlike anything ever discovered in science. It is much more like Columbus discovering another world than like Newton discovering gravity. Because, in effect, that is exactly what physicists have done, they have literally discovered an*other* world, an*other* domain, an*other* sphere of influence. Yet this new world is less like a foreign landscape to be surveyed and more like a familiar dreamscape with which to be reacquainted.

---

[101] Though not necessarily the sacred or divine realm.

Be sure, the metaphysical implications of quantum reality were not lost on the first generation quantum pioneers—those such as physicist Sir Arthur Eddington, who said "The stuff of the world is mind-stuff"[102], and physicist Sir James Jeans who stated that "The universe begins to look more like a great thought than a great machine."[103]

Eddington and Jeans are by no means alone in their assessments. Recall Bohr's assertion that "There is no quantum world. There is only an abstract quantum description."[104] Of course, this "abstract quantum description" *is the quantum world!* Heisenberg explains,

> In the experiments about atomic events we have to do with things and facts, with phenomena that are just as real as any phenomena in daily life. But the atoms and the elementary particles themselves are not as real; they form a world of potentialities or possibilities rather than one of things or facts.[105]

What's more, 'observation' is the magical event which turns possibility into actuality. As John Wheeler famously put it:

> No elementary phenomenon is a real phenomenon until it is an observed phenomenon.[106]

Physical reality therefore, is thus observed (i.e. measured) into existence, brought into being out of pure ethereality. Again, as unbelievable as all this sounds it is orthodox 21st century physics—*not* mystic mumbo jumbo, *not* pseudo-scientific bunk. N. David Mermin affirms,

> There is no deep reality.... Everyday phenomena are themselves built not out of phenomena but out of an utterly

---

[102] Sir Arthur Eddington, *The Nature of the Physical World* (1928).
[103] James Jeans, *The Mysterious Universe* (1930). Jeans also rightly perceived that "From the intrinsic evidence of His creation, the Great Architect of the Universe now begins to appear as a pure mathematician."
[104] Niels Bohr, *The Philosophy of Niels Bohr*, Aage Peterson (1963).
[105] Werner Heisenberg, *Physics and Philosophy, The Revolution in Modern Science* (New York: Harper, 1962).
[106] John Wheeler, in Nick Herbert, *Quantum Reality* (1985).

different kind of being. Far from being a crank or minority position, "There is no deep reality" represents the prevailing doctrine of establishment physics."[107]

We now know that material reality has immaterial foundations. This "immateriality" (which serves as the ground of all materiality) is quite literally a type of noumenality. Which, in turn suggest that the whole of physical reality emerges *cerebrally* (most likely, from the divine Mind).

Much is made of the noumenal character of quantum wave-function (and rightly so) but we must not forget that wave-function is but one side of the dualistic quantum coin.[108] For, out of noumenality, issues physicality—from particles, to planets, to people with minds of their own to contemplate the underwriting Mind of the universe!

This fact makes quantum wave-function all the more mysterious— *and all the more theistically suggestive.* For, inasmuch as quantum wave-function is the ground of the *physical* being of the universe, we are compelled to ask: what, in turn, is the ground of the *noumenal* being of quantum wave-function? The answer can only be God. As mind gives rise to consciousness, and hence, dream realities, so divine Mind

---

[107] N. David Mermin, in Nick Herbert, *Quantum Reality* (1985).

[108] Quantum event outcomes are physical manifestations. Science thus deems them to be naturalistic events.
These "outcomes" however derive from *preter*natural causes. This is of course possible because there are two sides of quantum reality: one naturalistic, the other *preter*naturalistic (that is, one physical the other *meta*physical). On the natural/physical side of quantum reality (i.e., our side) quantum event outcomes are physical (and hence natural) events, and as such are directly open to science. However, on the metaphysical wave-function side of quantum reality (i.e., God's side), quantum phenomena is not directly open to science, in that the phenomena is itself transcendent. From the wave-function side of quantum reality God is able to causally interface with the particulate constituents of the natural/physical side of quantum reality. God then, when acting at the quantum level, is not acting as a cause *within* nature. For, according to the Copenhagen interpretation quantum events have no *natural efficient causes*! Rather, the dualistic quantum ontology, insofar as it bridges both the natural world and the preternatural world, allows God to act as a 'transcendent' (and thus '*preter*natural') cause within the natural world, albeit in a manner intrinsically veiled to the prying eyes of science. The ontological structure of quantum dualism provides an utterly ingenious way for a personal, albeit transcendent God to act inside the otherwise naturalistic (and hence causally-closed) nexus of the physical world. God's quantum actions then, while preternatural, are both wholly objective and wholly within the phenomenological framework of quantum process—God's quantum actions then, are well within the bounds of scientific law. The sheer cleverness of the quantum ontic for cloaking the divine hand is as profound as it is telling.

gives rise to wave-function and physical realities.[109] What else could possibly be the source of the objectively existent, abstract, transcendent, noumenal, indeed platonic, substratum of the physical cosmos? The answers are, shall we say, greatly limited.[110]

Quantum reality straddles two worlds, plain and simple. The physical world of the here-and-now and the metaphysical world of the *there-and-after*.[111]

Science has unwittingly stumbled upon the "other" reality. A reality long intuited to exist by mystics and other spiritually sensitive persons. A general knowledge of this "other" exists implicitly within all persons, but the particulars have been made known revelationally by God through the Hebrew prophets, the New Testament apostles, and the only begotten Son, King Jesus the Christ. By and through (1) the inner witness of man, (2) the inscripturated Word of God, and (3) the ontological insights of quantum physics, mankind has come to the realization that material reality isn't the only reality. In fact, the materiality of commonplace reality would appear to be (substantively speaking) of a secondary, or lesser order, in that it actuates into physical being from an altogether different, more fundamental medium of being—one that is not only *not* physical, but actually *meta*physical!

Quantum dualism is scientific orthodoxy.

Matter/spirit dualism is theological orthodoxy.

Both agree, reality is dualistic; and the ontological recipe is physicality/metaphysicality wherein the latter has existential primacy over the former.

Scripture informs us that reality is material—but also spiritual. We read of physical bodies—and of metaphysical souls. We are told of that

---

[109] An imperfect analogy to be sure.
[110] Notwithstanding solipsism, the clearest answer seems to be God.
[111] A few hundred years ago academicians would have considered quantum phenomena 'magic' and *not* science, (just as many early twentieth century physicists had). However, unlike when primitive people attributed supernatural activity to what would later be shown to be mundane natural phenomenon (thunder-unto-Thor for instance), the genuinely preternatural character of quantum reality will never be explained away, for it is an ontological fact of reality. Here is a mystery that cannot be made mundane by simply explaining how it works. For, explaining how it works brings you face to face with the mystery itself! (cf. Feynman *Lectures on Physics* vol. 3. 1965).

which is sacred—and that which is profane. We are told of the holy—and the unholy. We are told of the divine—and the mundane. We are told in no uncertain terms of realities that transcend the ordinary, every-day dimensions of the here-and-now.

Reality then, as presented in Scripture is dualistic, with a corporeal realm and an incorporeal realm existing in transcendent/immanent communion. This is perhaps the most basic, most elemental metaphysical claim of Scripture, and quantum physics bears it out in full! To the chagrin of material monists, quantum mechanics has brilliantly corroborated the ontological dualism that is both predicted and presumed in the Judeo-Christian Scriptures.

If Christianity is true, as believers contend, then quantum dualism and the Christianity-pandering phenomena associated therewith isn't something that believers should necessarily be surprised at. For, if Christianity is true, the world *has to be this way*—or, some way very similar. As I have tried to show, a quantum, or 'quantum-like' ontology is *actually necessary* for a Christian realist reality (such as the one we experience) to exist.[112]

---

[112] Prior to the quantum age, the notion of a "Christian ontology" (that is, an ontology capable of realizing the metaphysical demands of the Christian faith) would have been considered impossibly unlikely. Such an ontology would be so scientifically unthinkable in its logical and rational workings as to be intelligibly unfathomable. It is on this very point that our universe gets interesting. For, contrary to the anti-supernatural biases of liberal Christianity and militant atheism, *the scientifically unfathomable claims of Christianity turn out to be phenomenologically indistinguishable from the scientifically unfathomable claims of quantum mechanics*! (Whether it be the fact that neither particular divine acts nor quantum events trace back to natural causes; or the reality of an objectively existent 'noumenal transcendence'; or the shadowy reflections of omnipresence, omniscience and/or middle-knowledge cast by quantum wave-functions, et al.,). This is truly a remarkable revelation from reality. I am not here speaking simply to the fact that Christianity and quantum physics are both scientifically unthinkable; but rather to the fact that the scientifically unthinkable ontology necessary to a Christian realist reality *is in fact realized in, by, and through the ontology of quantum physics*! As it happens, virtually all the same rational absurdities and scientific impossibilities common to Christianity (and ridiculed so mercilessly by atheistic materialists) are phenomenologically commonplace at the quantum level! Why is it then, that we never hear militant atheists ridicule quantum physicists for believing and teaching the reality of events every bit as supernaturally tinged as those of Scripture? Whatever the excuse, one thing is sure: it is no longer fair for atheistic materialists to point to supernatural events in Scripture and say "that's not how the world works", because as astonishing as it may seem, that's exactly *'how the world works'*; this, according to quantum physics, the most accurate scientific theory ever known to science. Ergo my contention that the quantum ontology has been phenomenologically structured to realize scientific impossibilities of the exact sort necessary to underwrite a Christian realist reality. This is quite a creative feat. Consider: God created a highly rational, highly

Any who might doubt this need only consider the devastating impact that Laplacian determinism and the Newtonian ontology had upon orthodox Christianity during the 18th and 19th centuries.

Contrariwise, quantum ontology elegantly and economically fulfills the precise ontology both *predicted from* Scripture, and *demanded by* Scripture!

Glimpses of Eternity

We have thus far learned that: We live in a *uni*verse governed by two distinct realities. These distinct realities are governed by distinct set of laws. One set is lawfully rational, the other set is lawfully paradoxical.[113] One reality is finite and physical, the other reality is eternal and metaphysical. I speak of course, of classical (macro) reality and quantum (micro) reality.

Classical reality is governed by physical laws and processes. Quantum reality is governed by metaphysical laws and processes. What's more, it is the latter which gives rise to the former!

It now appears that the material reductionist assertion that "matter precedes mind" has been turned on its head—indeed inverted, such that the incorporeal both precedes the corporeal *and gives rise to the corporeal*. This changes everything. The mind reels. Could this uniquely theological dualism be suggesting something of the world to come? Does the fact that the metaphysical precedes the physical provide us with an ontic insight into the psycho-physical nature of the spiritual bodies of the eschaton? Or is it suggesting an even more ultimate theological truth. As already alluded to, in light of the uniquely dualist character of the quantum ontic (wherein the incorporeal both precedes and gives to the corporeal) I think it highly plausible that God, as the theologically

---

intelligible *classical reality*, and grounded it upon a highly irrational, highly paradoxical *quantum ontology*, but did so in a way wherein the preternatural processes of *quantum reality*, though situated within the larger overall framework of *classical reality*, never once steps on the rational intelligibility that is the hallmark of classical science! This is to me, one of, if not *the* greatest of God's creative feats, reflecting ingenuity, elegance, economy and utility, all in equal measure. This, is to me a strong indicator that the Judeo-Christian God not only exists, but is the Creator or the world.

[113] Quantum mechanics, (while paradoxical within a classical setting), admits its own context, being governed by laws and phenomenology different than we are used to. This is not surprising, for, these laws and phenomenology are the governing rule of immaterial (indeed metaphysical) systems.

defined "ground of all that is", is in one way or another simply cogitating the universe into being via the underwriting laws of quantum reality.

There can be no way to prove this of course.

Even so, the circumstantial evidence from the quantum ontology is remarkably strong.

Beneath Physicality

That a realm and ontology such as we discover at the quantum level even exists is highly suggestive. For, among other things, quantum reality is quite suited to serve as an information-based communications system between the *there* and the *here*; between the spiritual and the natural. Again, that such a system exists is exceedingly suggestive. Why should there exist the means for a disembodied will (or Will) to act efficaciously in the world if no such disembodied Will exists? Why the ingenious communications network if no ingenious communicator? Quantum reality, by its very nature, character, and structure, begs the existence of God. And does so not by what we do not know, but by what we do know!

And what we do know is this: beneath the macro-sphere of ordinary physical reality lies an abstract micro-sphere governed by a radically different set of laws and processes. These self-same laws and processes are uniquely suited (phenomenologically speaking) to serve as a mechanism for a transcendent Mind intent upon acting causally (albeit clandestinely) upon the physical constituents of the material world. Of course, this "transcendent Mind" would have to be God, seeing that this self-same "Transcendence" is the very one responsible for designing the system in the first place!

Beneath and Beyond

The physicality of the cosmos is an emergent phenomenon distilled from the transient fluctuations of countless virtual particles oscillating in and out of existence from the indefinable location that is the quantum void. Space itself is a quantum foment; a seething hive of virtual particles (or quantum vacuum fluctuations), such that matter, energy, and now spacetime itself all appear to be derivative ultimately of wave-function! In a nutshell then, physical reality emerges from metaphysical platonic reality, which in turn emerges from divine reality—i.e. the Mind of God—which is the ultimate reality.

One might say that the cosmos, in all its material grandeur issues forth (seemingly *ex-nihilo*) from the void of quantum nothingness. However, 'quantum nothingness' is not the same as 'absolute nothingness', for the former, unlike the latter, is rife with platonic existents—including *wave-functions*. The "quantum void" then, is more accurately a void of "physicality."

Here, we again come face to face with that mysterious transcendence (or 'quantum medium') from whence the whole of physical reality issues into being. This otherworldly substrate, whatever it may be, is responsible for the moment-by-moment continuance of the physical cosmos—a job otherwise attributed to the *Logos*.[114]

While many will be baffled by the greater implications of all this, the basic facts of the matter remain unambiguous. The most basic, being that the bottommost layer of physical reality *is not physical at all*, but is rather a mind-like meta-substance, that, while incorporeal, nonetheless serves as the matrix of all physicality. This fundamental ontological fact is an empirically proven scientific fact, which abides against modern materialist sciences inherent (albeit archaic) Enlightenment bias against substance dualism, and/or anything that might even subtly speak to the supernatural.

I suppose that just as the geocentrist academicians of the 17th century had to learn that a scientific fact is a scientific fact regardless of how metaphysically unappealing (or appealing) it may be; and just as Victorian believers had to get used to the idea of biological evolution; so today's scientistic atheists need to get used to the idea of metaphysical reality. For, this is exactly what quantum physics has revealed: the existence of a "metaphysical reality."

The implications here are profound. Particularly as regards brain science and philosophy of mind. Contemporary (reductionist) science

---

[114] I am not of course implying that the quantum void is itself *Logos*; but rather, only emphasizing that the physical universe originates from that which is non-physical, and indeed noumenal. I ask: is this not exactly what we would expect to discover if the universe were in fact a free creation of God?—i.e. a physical reality given by and from a spiritual reality?

contends that "mind" is nothing more than an epiphenomenon—a mere by-product of the brains physical processes. According to this view psychological process and neurophysiological process are one and the same, rendering mind an illusion of chemistry.[115]

The epiphenomenal view of mind is reminiscent of the view of Pierre Cabanis, who, two centuries ago famously stated that "brain secretes thought like liver secretes bile."[116] More recently, John Searle has similarly stated that "mental states and processes are real biological phenomena in the world, as real as digestion, photosynthesis, lactation or the secretion of bile."[117]

Such views are common among contemporary physicalist scientists, who, being material monists, have no place in their philosophy for anything that might even remotely resemble a dualist conception of soul or spirit. Unto these, cognitive function *is* brain function; nothing more, nothing less. For this and other ideological reasons, these are beholden to the notion that matter precedes mind. The problem these have is that quantum physics tells us just the opposite!—*that is*, that mind (i.e., as reflected in objectively existent abstract mental constructs) precedes matter. What's more, it is fundamental, being *ontologically prior* to all physicality!

Physical reductionism then, along with its presumption of material monism is in stark disagreement with the empirical findings of contemporary physics—according to which there exists a mind-like essence within a mind-like domain that precedes the whole of physical reality—matter, energy, space and time.

The notion that there exists a spiritual realm *distinct from* and *prior to* the physical realm is no longer metaphysical conjecture but scientific

---

[115] While the majority of contemporary neuroscientists and philosophers of mind attempt to understand consciousness in terms of purely naturalistic phenomena a minority of others do not. David Chalmers for instance believes that consciousness may be ontologically fundamental (see Chalmers, *The Conscious Mind: In Search of a Fundamental Theory*, Oxford: Oxford University Press [1996]). If the Copenhagen interpretation of quantum theory is correct (And it is correct [cf. Aspect, 1982]) then something akin to what Chalmers has in mind cannot be far from the truth. It is certainly closer than anything dreamt of in the reductionist philosophy.

[116] This famous line originally came from Pierre Cabanis (1757-1808), *Relations of the Physical and the Moral in Man*, (1802).

[117] John R. Searle (1987). Searle has made this claim repeatedly.

fact! Of course, this is what Christianity has long taught.[118] Namely, that the physical arose from the spiritual—from the mind-like—from the *Logos*. And did so a finite time ago. And what's more, *continues to do so*, moment-by-moment, instant-by-instant, such that all depends upon the faithfulness of God as Creator and Sustainer of all that is!

I ask then, how can so many still be of the erroneous Enlightenment opinion that science has somehow disproved the existence of God? This is not only far from the case, *it is opposite the case!* For, quantum theory has ontologically (and hence, philosophically) reversed the Enlightenment worldview (which, recall, was monistic, materialistic, deterministic, reductionistic, and atheistic). This being the case, I imagine that if Napoleon and Laplace were around today, their conversation would have went *a lot* differently!

Quantum theory has to be the most far-reaching scientific discovery ever made. For, it has taken us behind the scenes of ordinary reality and put us into contact with the reality behind physical reality. This "reality behind reality" is a veiled reality. And the veil is physicality itself. Quantum physics has thus revealed the existence of an unseen primary reality that is not only non-physical but *meta*physical, yet every bit as real as the physical. This "primary reality" is an objectively existent reality that is abstract, eternal, and strikingly similar to that intuited by Plato over two millennia ago.

Ironically, mankind's first scientific encounter with a truly metaphysical domain has come about through the hardnosed discipline of physics. Consequently, science has been forced to broaden its definition as to what constitutes objective reality.

Physicists have literally been forced to recognize the existence of an ontological overlap; that is, an intersecting domain where the physical and metaphysical meet, and in fact, commune. This is of enormous significance. For, here is evidence of the long sought intermedium, or locus of interfaction, where the physical and the spiritual come into contact with one another for to communicate. And this, according to

---

[118] The existence of a spiritual realm is not unique to Christianity, of course. However, an incorporeal realm that exists transcendently and immanently both beyond and throughout the physical realm, with an ontological structure that allows a disembodied agent to act objectively and clandestinely in the physical world, over/against certainty of knowledge by free willed beings is; *and necessarily so.*

quantum physics, is exactly what they are doing! *They are communicating information.*[119]

Understand that this "blurring of, and between, realities" is not merely epistemic but ontologic. In other words, science can no longer determine (with absolute certainty) where physical reality ends and metaphysical reality begins! (Though science is not without good approximations).

Quantum dualism is the heart of this communion. And in it we discover a unification of sorts. A unification of what Augustine deemed to be "God's two books": the *Book of Nature* and the *Book of Scripture*—tidily bound together into a unified Tome by the ontic glue of quantum dualism, which allows the physical and the metaphysical to commune in a natural and efficacious way that preserves the integrity of both!

Understand, this is not a discovery to be taken lightly. For, it is none other than the long sought threshold between realities—*the elusive Cartesian bridge between worlds!*

This bridge however is far more specialized than anyone could have imagined, revealing a robust ontological interdependence between quantum physics and Christian theology.

Quantum ontology realizes Christian doctrine, and Christian doctrine necessitates quantum ontology.

Unfortunately, (but not at all surprisingly) these profound existential truths are lost on the majority of mainstream physicists, many of whom rarely (if ever) pause to consider the greater metaphysical implications of quantum theory—much less the greater theological implications of quantum theory. For most, the issue of quantum metaphysics ends with either Schrodinger's cat or Wigner's friend. Again, this is not at all surprising, given the standard, secular "shut-up-and-calculate" approach to physics in the west (which is rigidly pragmatic, and seemingly obsessed with technological advance).

Fortunately, not all physicists are so forward thinking that they cannot see what lies just behind the quantum veil.

There are prominent physicists at work today, in this early 21st century (e.g., Robert John Russell (1946-), John Polkinghorne (1930-),

---

[119] Prior to, and during an ordinary quantum event, a physical object temporarily ceases from physical existence. In its stead, is only abstract information regarding the objects properties and future states. From this information outcomes are indeterministically made and physical re-actuation occurs, thus ending the process.

et al.,), just as there were in the early-to-mid decades of the twentieth century, when quantum theory first came to light (e.g., Arthur Compton (1892-1962), William Pollard (1911-1989), Karl Heim (1874-1958), Eric Mascall (1905-1993), et al.,) who instinctively recognize in quantum theory deeper implications for issues of theology.

For mainstream physicists, quantum mechanics is as a wellspring for technological innovation. For those with eyes to see, it's a hotbed of theological insight. Again, these insights were not lost on the early twentieth century quantum pioneers. For instance, almost as soon as Bohr and Heisenberg formalized their theory (1927), Arthur Holly Compton advanced his indeterminacy-based model of human free will (1931).[120] Karl Heim, E. L. Mascall, and William Pollard followed shortly thereafter with indeterminacy-based models of divine action.[121]

Insights of this sort tend to be lost on contemporary physicists—who, having been trained to work wearing secular pragmatist spectacles, have no vision for metaphysics. These can tell you exactly how wave-functions evolve, and exactly how wave-functions collapse, but have absolutely no idea what 'wave-function propagation' has to do with the theological notion of 'divine middle-knowledge.' (And why should they? Physicists, after all, are not [generally speaking] theologians. By the same token, theologians are not [generally speaking] physicists. Ergo the interdisciplinary nature of quantum-theological issues. Not to mention its lack of mainstream popularity).

It is quite possible then, for a physicist with an acute knowledge of quantum physics to be completely in the dark with respect to the ontological implications of quantum physics for Christian theology!

---

[120] In Compton's view (which I believe to be correct) human will is the hidden quantum variable. Consider Compton's account (as given in a 1955 *Atlantic Monthly* article); "A set of known physical conditions is not adequate to specify precisely what a forthcoming event will be. These conditions, insofar as they can be known, define instead a range of possible events from among which some particular event will occur. When one exercises freedom, by his act of choice he is himself adding a factor not supplied by the physical conditions and is thus himself determining what will occur. That he does so is known only to the person himself. From the outside one can see in his act only the working of physical law. It is the inner knowledge that he is in fact doing what he intends to do that tells the actor himself that he is free." See, Johnston, ed., *The Cosmos of Arthur Holly Compton*, (New York: Knopf, 1967).

[121] See, Heim, *The transformation of the scientific world view* (London: SCM, 1953). Mascall, *Christian Theology and Natural Science: Some Questions in Their Relations*, The Bampton Lectures, 1956 (London: Longmans, 1956). And Pollard, *Chance and Providence: God's Action in a World Governed by Scientific Law* (London: Faber and Faber, 1958).

(And the same goes for the theologian with no knowledge of quantum physics.).

Even so, no orthodox physicist would deny the reality of 'quantum dualism'; or the fact that wave-functions are 'abstract' in addition to being 'non-physical.' *Nor would they deny that wave-function 'abstractness' can and does affect physical matter (such as the location of individual particles in single-particle double-slit experiments). Or that 'wave-function itself' can be affected by physical matter (as in quantum collapse).* These latter examples are clear manifestations of *metaphysical-to-physical* and *physical-to-metaphysical* interfaction! And yet remarkably, they aren't even recognized as such by modern science, because modern science doesn't have a category for *"metaphysical phenomenon!"* *This, despite the fact that quantum theory is shot through with metaphysical phenomenon!*

Recall, the M/O for the physicist is to shut-up, calculate, advance technology, and leave the interpretation of meaning to the philosopher, metaphysician, and theologian. The fear, I suppose, is that science will get stuck in a metaphysical bog. Such, at least, seems to be the anxiety of Feynman's famous admonition to physics students,

> I think I can safely say that nobody understands quantum mechanics… Do not keep saying to yourself, if you can possibly avoid it, 'But how can it be like that.' Because you will get 'down the drain', into a blind alley from which nobody has yet escaped. Nobody knows how it can be like that.[122]

From here, it is a very short distance to "shut-up and calculate".

Feynman is absolutely correct. Any physicist attempting to fathom "how" or "why" quantum physics is the way it is, is sure to end up down a blind alley. The reason being, that the answer to this mystery doesn't lie in physics—*but in theology!* Accordingly, one wearing only the spectacles of a physicist will surely be blind. Little surprise that Pollard, Heim, Russell, Polkinghorne, et al., all have training in both physics and theology. It would seem that such is required for insights beyond what either, in themselves, could tell us. One might say (after

---

[122] Feynman, *The Character of Physical Law*, (BBC/Penguin, 1965).

the spirit of Einstein) that "religion without quantum physics is lame, and quantum physics without religion is blind."

Within from Without

As we have seen, the laws and processes of quantum theory are transcendent, existing beyond and/or outside of the spacetime dimensions of this material world. This is interesting insofar as theologians as early as Augustine (AD 354-430) have posited that God is transcendent, existing beyond and/or outside of the spacetime confines of this material world—within a higher plane, as it were.

This is a quite rational sentiment. Logically speaking, if God created the spacetime continuum that is the cosmos, then He presumably did so (relatively speaking) from a point or locale external to the continuum being created. For this reason, God is thought to be not only immanent within and throughout creation, but also transcendent above and beyond creation—and so also, *separate from* creation.

This being the case, if I am correct in my assertion that the quantum wave-function serves as a sort of 'divine-to-physical' upload mechanism that converts the divine intentions of God into seemingly naturalistic events, then it comes as no surprise that we find this unique 'preternatural-to-natural' causal converter mechanism existing within the same sphere of reality that theologians have traditionally located God—which, is of course, just beyond the periphery of this physical reality![123]

Mechanisms of Creation and *Creation Continua*

Big Bang cosmology has long posited that the universe came into existence from nothing—i.e., from where neither matter, energy, space, nor time exist. Along these lines, quantum cosmology also posits a universe from nothing—or more specifically, *nothing that is physical.* According to quantum cosmology the universe 'inflated' into being via a quantum vacuum fluctuation—which, in turn, emerged out of

---

[123] Let me be clear on this point; I do not mean to suggest that higher realities, such as say, higher dimensions are themselves supernatural, only that quantum reality, being *both* noumenal, transcendent, and immanent, fits incredibly well with my overall postulate that quantum mechanics is divinely ordained to serve larger divine purposes.

wave-function. According to this view, the universe arose from processes that are both ontologically beyond and prior-to itself. Processes that are by definition transcendent—not to mention abstract.

Again, what we thus have, is a realm that is abstract, transcendent, and "prior to" space and time as we know it. Again, what we thus have is a realm eternal.[124] A realm not unlike that imagined in platonic thought.[125]

In a nutshell then, physical reality emerged from a transpatiotemporal reality—that is, "a wave-function"—*behind which* was God (the *Logos*), and *into which* God (the *Logos*) *breathed fire*, creating a universe where before was only mathematical abstraction!

What this ultimately means is that the corporeal world initially issued forth from incorporeality, and subsequently continues to issue forth from incorporeality. Quantum wave-function then, is both the *province* and *provenance* of creation!

The unification of the physical with the metaphysical is reflected in the dualistic nature of the quantum wave-function, which seems to be governed by an open process of reciprocal creativity wherein *potential physical existents* (i.e., potential physical states defined by and within the abstract wave-function) give rise to *actual physical existents* (i.e., quantum particles) that again give way to *potential physical existents*, and on and on, physical particles appearing and disappearing, ever

---

[124] Philosophers of time speak of "eternity" in two distinct ways; One classical, the other modern. In the classical conception, eternity is atemporal and timeless (i.e., without duration). In the modern conception, eternity is temporal and everlasting (i.e., with duration). According to the first of these, eternity is a static, "ever-present now" so to speak. Herein exist those static, timeless, abstract objects of Platonic thought, such as the geometric shapes (triangles, circles, squares, etc.,), the natural numbers (1, 2, 3, etc.,), and the like. Conversely, according to the second of these, eternity is a temporal, everlasting, duration of time. Herein exist those dynamic abstract entities of platonic thought, such as stochastically evolving wave-functions, and the like. So then, quite apart from the question of God's relation to time and eternity, it is obvious to me, that both static time and fluid time exist within realms metaphysical, in, by, and through abstract objects. What's more, these realms, insofar as they take up abstracta, have their ultimate grounding in the divine Mind, possibly within a conceptualist mode of being. It thus seems that there is (1) physical time (or creaturely time, such as we experience), of which God is outside. As well as (2) metaphysical time (such as the (a) static and (b) fluid times associated with Platonic/quantum entities), of which God is also outside. And finally (3) God's time, which, while "outside of" (and/or beyond) all others, nonetheless encompasses each.

[125] Whether or not the quantum cosmological view of origins is correct, the fact remains that the universe, having a finite existence, came into physical being from physical non-being. This fact is to me theistically irresistible.

shifting between physical existence and mere potential existence as the mathematical abstraction that *is* quantum wave-function interacts with physical reality. The onto-existential dynamic here is wave-function evolution, followed by wave-function collapse, followed by wave-function evolution again, ad infinitum—or, until the eschaton when the heavens and earth are made anew.

This repeating process of wave-function evolution/collapse/evolution reflects a sort of onto-existential reciprocity between *physical* reality (i.e., actuation [or, wave-function collapse]) and *potential* reality (i.e., wave-function propagation), with wave-function serving as a sort of 'expectation medium', between "what is" and "what may be." From a theological perspective, it might be said that wave-functions provide localized sets of middle-knowledge[126] from whence divine determinations are made.

Middle-knowledge is an intrinsically platonistic concept. As such, it fits nicely within the quantum framework, which is a unique expression of platonic metaphysics. Consider: quantum reality speaks to the existence of both abstract entities and physical entities—wherein the former gives rise to the latter. And herein is what makes quantum reality platonistic. For, logically speaking, any entity that gives existence to another entity must (*necessarily*) be at least *as real* as that entity. Wave-functions then, though purely abstract and mathematically structured are every bit *as real* as the physical entities they give rise to! I, in fact, posit that wave-functions are even more real than the physical constituents of the universe insofar as wave-functions are transcendent, mathematical, and eternal. Mathematical ideals after all are sublime

---

[126] Middle-knowledge is an aspect of God's omniscience recognized by theologians since the 16th century. Originally posited by Jesuit Scholastic Luis de Molina (1535-1600), God's middle-knowledge includes: (1) divine knowledge of everything we will do; (2) divine knowledge of everything we would do in every possible situation and under every possible circumstance (an early theological version of modal realism); and (3) divine knowledge of what persons who do not exist would do in every possible situation and under every possible circumstance, had they existed. God then, can not only see all that happens but also all that may (or may not) happen. God's middle-knowledge-based perspective of reality (including His knowledge of every potential outcome of every past, present and future quantum event) would provide him with ultra-Laplacian insight into the virtual infinity of possible quantum state outcomes (as well as each, over and against every other). God then, would have at His disposal countless (*potentially existent*) quantum-induced macro-level outcomes. As such, the presumed 'myriad' of quantum events thought to be necessary to precipitate a single macro-level event would not be required; rather, only a single "right" quantum event placed at a single "right" moment in history.

realities not given to temporal corruption—physical realities however, are. We thus have 'physicality' arising from that which is abstract and ideal, not to mention *noumenal!* And possibly even *"Noumenal!"* Simply put, physical reality owes its origin and continuation to an abstract reality *more real* than itself! A reality that is more like consciousness than anything else we know of.

Quantum physics confirms what Plato deduced more than two millennia ago: that physical reality is a shadow reality—a veil, behind which ultimate reality—i.e., divine reality, lies. As odd as it may sound, physical reality comes *directly into being* from mathematical reality. Little wonder so many quantum pioneers had platonist elements in their thinking—e.g., Werner Heisenberg, Niels Bohr, Wolfgang Pauli, Max Born, Max Planck, Albert Einstein, Erwin Schrödinger, Eugene Wigner, Paul Dirac, John von Neumann, et al.[127]

As already mentioned, quantum dualism is not the exact "Platonism" of classical thought (cf. Plato, Plotinus or Proclus) but it is nonetheless "platonistic" in that the physical derives of the metaphysical. What's more, in this physical reality, *wave-function* is the abstract 'Ideal'. And it keeps physical reality percolating into being, particle-by-particle, statistical-value-by-statistical-value, just as the above-mentioned collapse/actuation cycle dictates.

This metaphysical-to-physical cycle is driven by interactions that take place between propagating wave-functions and classical-sized matter aggregates—e.g. tables, chairs, photocells, etc.,

Interactions and/or measurements collapse wave-functions (termed technically as the 'reduction of the state vector') such that all middle-knowledge (i.e., potential actuates, technically termed 'eigenvalues') present within the system up to that point instantaneously vanish, apart from one, which, in that same instantaneous moment actuates

---

[127] Many of the great modern mathematicians were (and are) mathematical realists/platonists. From the great G.H. Hardy, who said: "For me, and I suppose for most mathematicians, there is another reality, which I call 'mathematical reality'; and there is no sort of agreement about the nature of mathematical reality among either mathematicians or philosophers ... I believe that mathematical reality lies outside us, that our function is to discover or observe it, and that the theorems which we prove, and which we describe grandiloquently as our 'creations' are simply our notes of our observations." (G.H. Hardy, 1940); to Alain Connes (1982 Fields Medal winner), who said: "Prime numbers ..., as far as I'm concerned, constitute a more stable reality than the material reality that surrounds us. The working mathematician can be likened to an explorer who sets out to discover the world." (Changeux and Connes, 1989).

into a physical constituent of reality—be it an electron, photon, atom, molecule, or other.

Wave-function collapse is both instantaneous and discontinuous. Yet, in that mysterious instant, a solitary act of creation takes place as an abstract potentiality (i.e., a mere mathematical value) gives place to a physical actuality (i.e., a constituent of the cosmos). Within this discontinuous instant, the wave-function, which is merely the abstract probability of existence, ceases from being, and actual existence at once begins—and does so, as it were, out of nothing!

In this discontinuous instant something quite original happens; probable existence gives way to actual existence. And herein lies a mystery for the ages. How does a mere mathematical value, i.e., a mere probability of existence, give place to actual existence? How does an abstract *probability of* existence become an *actuality in* existence?

To this question, science has no answer. And rightly so. For, within this self-same instant, a threshold is crossed—viz., that separating the physical from the metaphysical.

At this point in the quantum account science hits an explanatory wall, but rather than look to other disciplines for insight (e.g. theology), science chooses rather to makes a declaration of "brute fact" and be done with it (as if the issue had been resolved *ex officio*).

What's more, contrary to certain scientific conceits, the end of "scientific" explanation is not the end of *all* explanation (as if scientific knowledge were the only legitimate knowledge). Indeed, the very success of science itself requires an explanation beyond that which 'science itself' is equipped to answer! Take the intelligibility of the cosmos for instance; while it can be well discerned by science, it cannot be explained by science. For *that* kind of understanding, you need theology.

You see, "discovery" and "explanation" are two very separate things. And what science 'discovers', it often cannot also 'explain.' And why should it, after all, the objective of science is to tell us "how", not "why." The latter is a metaphysical question, and is best served through metaphysical ends.

It is important to understand that not everything can be (nor is meant to be) "naturally" understood. Where scientific inquiry fails, theological insight can be invaluable. Simply consider the "quantum event" threshold spoken of above: where science leaves off (explanatorily), theology picks up (explanatorily), and does so beautifully via the long held doctrines of divine middle-knowledge and creation continua!

Consider for a moment just how amazing this realization is: Science, working at the foundational level of reality, comes to an impasse; an explanatory wall. With nowhere else to turn, theology steps into the situation with perfectly sculptured explanations that, though centuries old, are somehow flawlessly fitted to the very enigmas just realized to exist through quantum physics!

Returning to the above discussion regarding the nature of quantum collapse, the discontinuous shift from abstract potentiality to manifest physicality occurs the instant a wave-function, engaged by a classical system, produces a measurement.

For instance, the wave-function of say, an electron will shift from potential existence to actual existence the instant said wave-function impinges upon an atom in a macroscopic system—such as a phosphor detection screen.

In a discontinuous a-temporal instant, impingement precipitates an act of *creation continua* as a physical particle leaps into existence from what was an instant before a mere "potentia" existing solely as a mathematical reality; a statistical value; a solution to the Schrödinger equation—i.e., a wave-function, possessing only a noumenal existence within the platonic sphere of mathematical realities, ideals, and/or forms.

Mysteries of the Quantum Event

From a scientific standpoint, quantum events (i.e., collapse/actuation outcomes) are a two-fold mystery.

(1) The first of these mysteries concerns the cause of quantum collapse: What is it exactly that "triggers" the collapse of a wave-function?
(2) The second of these mysteries concerns the cause of outcome determination: What is it exactly that "determines" the outcome of an actuation?[128]

---

[128] Quantum event outcomes are stochastically constrained, but insufficiently determined. Regarding the latter, a question arises with respect to the degree unto which God acts to determine quantum event outcomes. For instance, does God determine *all* quantum event

With regard to question one, physicists know that registry interactions (i.e., measurements) trigger wave-function collapse; but there is as yet no good reason why.

As regards question two, scientists are largely content to say simply that such outcomes are ontologically indeterminate. That is to say, *nothing* makes the determination. This is not a very satisfactory answer however, since we know that outcomes do in fact obtain! And herein lies a mystery. For, according to theory, nothing in nature determines the outcome, yet selections are made; determinations do appear!

Some might suggest that perhaps the answer lies 'outside of nature', such as within the wave-function itself. However, there is nothing in, or within the wave-function to account for 'outcome' phenomenon. Consider: how could an abstract wave-function possibly make a selection as to one particular outcome to the exclusion of all others? How might an abstract wave-function exhibit "preference", particularly when this "preference" has to be deliberate, and rational (for, to be in statistical compliance with mathematical rule?).

---

outcomes as Nancey Murphy believes? (see, Nancey Murphy, 1995, "Divine Action in the Natural Order: Buridan's Ass and Schrödinger's Cat", in *Chaos and Complexity*). Or does God determine only *some* quantum event outcomes as Thomas Tracy believes? (see, Thomas Tracy, 1995, "Particular Providence and the God of the Gaps", in *Chaos and Complexity*). Or does God perhaps utilize a combination of the two, as Robert John Russell believes? According to Russell God acts in all quantum events until the rise of life and consciousness, after which God limits His determinations. (See, Robert John Russell, 2008, *Cosmology from Alpha to Omega: The Creative Mutual Interaction of Theology and Science*). Murphy's view and Russell's view have their merits, however I believe Tracy's view provides the most accurate understanding of divine quantum action, for reasons having to do with (1) theodicy, (2) God's desire to avoid divine determinism, and (3) God's desire for a naturalistic realm that while governed by process regularities, is nonetheless endowed with an intrinsic degree of freedom-of-process, for to avoid our living in a rigidly deterministic ontology—divine or otherwise. On a separate note, there is the question of "how" God goes about determining quantum indeterminacies? This is a mystery, of course. However, we can speculate. For instance, if God upholds the whole of quantum reality, *potentiality waves and all*, (as I believe He does), then God has direct access to the worlds various wave-functions, and so can directly specify particular quantum event outcomes while perhaps loosely determining all others in accordance with their stochastic limitations. Whatever the reality may be, there are clear implications from quantum indeterminacy to the classical divine-sovereignty/free-will paradox. Thanks to the open character and structure of quantum indeterminacy, neither "agent-action" nor "Agent-action" are any longer in competition with natural causation! After the spirit of Wigner, it could be said that *the miracle of the appropriateness of quantum indeterminacy for the realization of Christian theology is a wonderful gift which we neither understand nor deserve.*

Again, to say simply that the quantum outcome is "indeterministic" is unsatisfying since the reality is that quantum selections do in fact obtain—and do so every time a wave-function collapses. Consider: for, where there is selection, there is also preference, and where there is preference, there is also volition—And where there are these things *there is an act of will!*

And of course, where there is will, there is also mind.[129] And this takes us directly to one of the central assertions of this work: namely, that the noumenal essence of wave-function is intended to be directly accessible to the Mind of God.

Consider; physical substrates trigger instantaneous wave-function collapse, however, non-physical substrates (such as other wave-functions) do not. Overlapping wave-functions for instance, can and do co-exist in superposition (as is known from double-slit experiments). Wave-functions then, can and do incorporeally engage one another without inducing collapse. So while physical substrates (such as macro-sized matter-aggregates) induce collapse necessarily, incorporeal substrates do not.

Incorporeal entities that aren't wave-function then, may do the same (i.e., engage wave-functions without inducing collapse). This would mean that an incorporeal entity, such as the Mind of God, might engage a particular wave-function, for, to determine a particular quantum outcome, albeit without necessarily inducing collapse!

Just as two distinct wave-functions can interact, being of the same noumenal essence, so the divine essence can interact with individual wave-functions for to "select out" preferent actuations in accordance with the divine will.

Whether or not wave-function is a divine essence, we cannot know. However, insofar as divine communication is concerned, it seems to be enough that wave-functions are statistical and noumenalistic. Either way, the quantum essence appears uniquely contrived to be in service to the divine Mind (and perhaps even to the human soul, or spirit).

---

[129] C.S. Lewis agrees that the mind behind physical reality is preferential. Lewis writes: "[I]t is conscious, *and has purposes, and prefers one thing to another*... it made the universe, partly for purposes we do not know, but partly, at any rate, in order to produce creatures like itself...to the extent of having minds." [Italics mine]. C.S. Lewis, *Mere Christianity* (New York: Macmillan, 1960).

Ψ and the Will of God

The collapse of the wave-function then, is in reality, the locus of a divine dynamical process wherein the divine will chooses for existence a single eigenvalue (i.e., potentiality) out from a possible multitude of alternatives. These "alternatives", as we have seen, exist on a platonic level, in the seeming capacity of divine middle-knowledge.[130]

---

[130] Critics of divine quantum action (DQA) have contended that Copenhagen interpretation-based DQA is in reality a hybrid between two opposing interpretations of quantum theory; namely the Copenhagen interpretation and the hidden variables interpretation, albeit with God as the hidden variable. (See M.J. Dodds, *Divine Action: Contemporary Science and Thomas Aquinas* (Washington D.C.: Catholic University Press of America, 2012). However, this critique only has teeth *if* proponents of DQA are positing God as a *naturalistic* determiner (or naturalistic hidden variable) of quantum events. Nobody however is positing this. In my own DQA model for instance I posit God as a *transcendent* determiner, not as a divine being acting as a *naturalistic* hidden variable. The critique then is specious. For, while hidden variable theories presume to invoke "naturalistic" hidden causes, the Copenhagen interpretation, (such as underwrites my own DQA theory) affirms that there are no naturalistic efficient causes for quantum event outcomes. Whatever the determiner of quantum events outcomes then, it must be preternatural. Such charges then, along with those that claim that DQA reduces God to a natural cause are simply erroneous. Robert John Russell elucidates; "[I]f science claims that there is no complete set of natural causes for a quantum event, then we can argue that the addition of divine causality brings these events to completion without violating these laws or without being equivalent to a natural or secondary cause." (R.J. Russell, "Religion and the Theories of Science: A Response to Barbour", *Zygon: Journal of Religion and Science*, Vol. 31, Issue 1, [1996]). And elsewhere: "God is not acting as a natural efficient cause." (Russell, 'Does "The God Who Acts" Really Act?' in *Cosmology: From Alpha to Omega; The Creative Mutual Interaction of Theology and Science* [Minneapolis: Fortress Press, 2008]). In my particular version of divine quantum action, God is a *transcendent* cause that is able to act in nature through the uniquely dualistic ontology of quantum physics, which, insofar as it bridges the finite-corporeal-realm and the infinite-incorporeal-realm, is able to serve as an interface between the mundane and the transcendent. I believe that the most adequate account of DQA will have to *necessarily* incorporate platonic realism into its understanding of wave-function ontology. Yet I do not hear anybody saying as much. The fact of the matter however, is that for God to interface with the natural world requires more than mere ontological indeterminism—it requires *platonic realism* as well! You see, indeterminism may provide ontological room for God to act in nature without violating any pre-existing natural laws, but it does nothing to help us fathom how the incorporeal might possibly interface with corporeal! This of course being the infamous "causal-joint" problem spoken of by Austin Farrer (Farrer, *Faith and Speculation: An Essay in Philosophical theology* [London: A. and C. Black, 1967]). However, once we realize that quantum indeterminacy is grounded in the uniquely platonic essence of wave-function (which is itself incorporeal as well as transcendent) we can at least begin to fathom how incorporeal-to-corporeal interfaction might take place. Ultimately, *indeterminism* and *wave-function realism* address two separate theological questions. For instance, while *indeterminism* helps us to fathom how God can act in nature without violating nature's laws, *wave-function*

This sphere of abstract potentialities serves not only to maintain the sustained openness of creations moment-by-moment continuance, but also as the means whereby the general providential hand of God actuates physical reality into being; particle-by-particle.

By upholding these foundational ontological processes, God continues the existence of all that is.[131]

In a nutshell, wave-functions are platonic islands of middle-knowledge with absolutely no *physical* standing in reality. Mysteriously, however, wave-functions possess an objective standing in reality prior to collapse! This standing, while non-physical, is nonetheless objective. Again, this is a mainstream fact of science, whether materialist scientists want to accept it or not. This tells us that *physical* reality is not the *only* reality. And that the rationality of the universe extends beyond mere physical structure.

There exists an abstract rationality that is completely independent of the universe itself. And it is this 'rationality' that quantum mechanics has put us into contact with. Here is an idealistic reality existing objectively in abstract proximity to the human mind!

This is mankind's first *actual* encounter with a ray of light from beyond the cave.

I say "first *actual* encounter" because up until the quantum revolution the existence of this "idealistic" reality was known solely through Platonic deduction. It is truly fascinating that Plato, living centuries before Christ, was able to deduce the existence of a whole "other" reality—one that modern science has only recently been made aware of.

Spiritually-minded scientists have long suspected the existence of just such a frontier, but with no way to know empirically, the matter was left to the philosophers, metaphysicians and theologians. Then, the quantum revolution happened, and everything changed. Materialism was disaffirmed. Causation was shown to have a blind spot. But worst of all, the line between mind and matter was irrevocably blurred. Physicists could no longer tell where their equations ended and the phenomenon began. The material world, and physicality itself, was

---

realism helps us to fathom how God (as Spirit) can act effectually in the material world. Quantum mechanics thus provides scientifically-inspired insights into age-old theological mysteries—insights that are both *profoundly elegant* and *elegantly profound.*

[131] By "all that is", I mean simply 'the physical cosmos', and not the divine realm.

shown to be utterly dependent upon abstraction—upon a quantum ideal—a "mathematicism." This 'mathematicism' is of course, wave-function. And for the first time in history, mankind *empirically* knows that physical reality is not the only reality! Nor, physical existence, the only type of existence!

That this 'idealistic' frontier exists objectively (but non-physically) is a scientific fact.

We know this because the particulate actuations that manifest upon wave-function collapse consistently match the mathematical predictions generated by physicists via the Schrödinger equation! This phenomenon, (like double-slit interference patterns), is only explicable if wave-functions have an objective (albeit non-physical) existence. Put differently, the success of the Schrodinger equation speaks to the accuracy of the mathematical model in approximating the reality.

We know that the Schrödinger equation closely approximates *the reality* because the solutions derived thereof are in striking agreement with experimental observation. The Schrödinger equation (which governs the wave-function) thus speaks to the objective existence of that unto which its approximations refer! This, of course, being the platonic wave ontology.

Quantum reality is a mathematically realist reality. Understand, this *is not* an opinion. This is simply 'the case.' This is just the way the world is. Plain and simple. Quantum physics then, for all its uncertainty, lets us know one thing for sure: nominalism is dead.[132] Abstract realities truly do objectively exist. In this truth, we must be extremely cautious. For, quantum metaphysics has already been radically misappropriated by the various postmodern gurus of the New-Age movement. In reality

---

[132] Nominalism (the assertion that abstract objects do not exist, but concrete objects do) had already been severely challenged by the Quine-Putnam Indispensiblity Thesis. Here is the Indispensibility Thesis as formulated by Mark Colyvan (1998): (P1) We ought to have ontological commitment to all and only the entities that are indispensable to our best scientific theories. (P2) Mathematical entities are indispensable to our best scientific theories. (C) We ought to have ontological commitment to mathematical entities. This is the Indispensability Thesis (IT) and it speaks to the ontological chauvinism of the nominalist. For, the same nominalist that rejects the IT with respect to the existence of mathematical entities, will at the same time accept the IT with respect to the existence of charm quarks and strange quarks! One cannot have it both ways! So, unless the nominalist is prepared to reject a great deal of science (including quantum physics!), they best come to terms with the objective reality of abstract/mathematical realities! Wave-functions in Hilbert space is a good place to start.

however, when and where quantum physics does get metaphysical, it gets metaphysical in highly particularized, highly constrained ways—ways usually directed in some way or another at the realization of a particular element of Christian reality.

It might thus be said that quantum physics has nothing to do with Eastern mystical assertions of mind-*over*-matter, and everything to do with Western Christian assertions of Mind-*before*-matter!

In the East vs. West battle for quantum mechanics, then, Christian theism is the clear winner. Recall that Christian theism is in fact so closely linked with quantum process that the former would actually be irrealizable in the absence of the latter!

The Quantum Substance

So what exactly is the mysterious quantum substance? This is a good question. Alas, the answer will probably forever elude us. However, this is not to say that we are without a clue. Far from it. All that we know of wave-function tells us that we are dealing with an essence that exists somewhere between 'platonic abstracta' and 'spirit substance.'

Consider the former: wave-functions are indeed "platonistic", for, they derive from the Schrödinger equation (which, recall, describes the deterministic propagation of an abstract, wave-like substrate). These however are more than mere mathematical abstractions in that they are capable of producing physical effects! Consider now the latter: Wave-functions are also "spiritualistic" in that they exhibit mind-like properties, as well as the ability to communicate information *non-locally!* Einstein referred to this ghostly phenomenon as "spooky action at a distance."[133]

---

[133] Quantum entities that physically interact become conjoined as a single non-separable system with a common wave-function. This phenomenon, known as "quantum entanglement" allows conjoined quantum systems to retain their connectedness over arbitrarily large distances (including many light years distance) resulting in instantaneous, non-local communication. This "spooky (or "ghostly") action at a distance" is a uniquely transcendent phenomenon that is (at least superficially) reminiscent of omniscience (insofar as entangled quanta have instantaneous knowledge of each other, even when separated by inestimable distances) and omnipresence (insofar as entangled quanta can instantaneously inform and effect each other, again, even when separated by inestimable distances). Quantum entanglement provides the means for instantaneous (non-local) wave-function causation.

Quantum physics has literally turned the materialistic, deterministic, reductionistic ontology of classical physics on its head by showing that reality is fundamentally immaterialistic, indeterministic, and non-reductionistic—exhibiting both holistic and non-local phenomenon.

It is not easy to overstate how truly revolutionary the quantum revolution is. Here is a revolution that is equally a revelation! A 'revelation' that the ultimate nature of reality is uniquely metaphysical—*and particularly theological*. As we have seen, quantum reality has two ontological sides; one physical, the other metaphysical (or perhaps more accurately, "spiritual"). It is not surprising then to discover that quantum physics plays fundamental phenomenological roles in both physical reality and spiritual reality—in service to both the corporeal body and the incorporeal soul.

Quantum reality is naked reality. Pure reality. Reality stripped of all physical appearances. Quantum reality defies mental construction. For, it appears to be mental *construct* itself! *Objectively existing construct!* The only question left, is whose construct is it? And from here, it is a small step to God.

The quantum essence, for all its uniquely mentalistic properties, is irresistibly spiritualistic, playing a constitutive role in the theological structure of reality itself. Whatever the quantum essence, it is eternal: preceding matter and energy, and transcending space and time. I suspect that this essence derives directly of divinity, but without itself being divine. As for purpose, all impression is that it serves as a bridge, medium, or go-between, for linking ontologically disjunct realms; namely, the spiritual and the material. Quantum phenomenology allows for just this, by providing a means whereby the incorporeal is able to commune and communicate with the corporeal (and vice versa). For, out from the quantum essence comes all that is, issuing forth as it were by divine decree, as God, the quantum Creator "determines" our world into being from beyond the quantum veil.

Chapter Five

The Mechanism of Divine Action

> It seems hard to sneak a look at God's cards. But that He plays dice and uses telepathic methods...is something that I cannot believe for a single moment.[134]
>
> —Einstein

> As I have said so many times, God doesn't play dice with the world.[135]
>
> —Einstein

> Einstein, stop telling God what to do.[136]
>
> —Bohr

Regarding the question of the mechanics of God's action in the world, it would seem that a small corner of the veil of the mystery has

---

[134] This particular statement comes from a letter to Cornel Lanczos, Mar 21, 1942; *Albert Einstein: The Human Side*, 1954.

[135] This famous quote speaks more to Einstein's deep-seated allegiance to deism than to anything else. Ironically, Einstein's deism blinded him from seeing the truth of ontological indeterminism. Not to mention the revelatory implications of ontological indeterminism for a theistic deity.

[136] In 1983 Alain Aspect experimentally proved Bohr right and Einstein wrong. God does indeed play dice. In fact, He loads them!

been lifted. Quantum physics has now placed the speculative question of "how God acts" on solid scientific ground by providing an empirically based system of mechanics wherein divine action might be said to occur.

This, of course, is not to suggest that science can in any way put the finger of God under a microscope, or the will of God in a test tube. Neither is it to suggest that science can in any way know qualitatively whether a particular action or event was brought about via a divine act or a natural act. Rather, it is only to say that we can for the first time in history fathom divine action in terms of well understood scientific process (and consider the modal character of that action in terms of its relevant scientific phenomenon).

Quantum mechanics offers the contemporary believer a (for lack of a better expression) 'conceptual impression' of God's causal interaction with the world in terms of the quasi-mechanical processes He utilizes at the quantum level. God's cards, held close to the chest, are difficult to see clearly. Even so, quantum mechanics offers a clear-enough meta-scientific account of how the long-sought-after God-to-world-interface so indispensable to the Christian world-view might possibly be realized in nature.[137]

Quantum phenomenology, like divine action, is an arcane topic. Neither however, is wholly beyond the cognitive grasp of the human mind. For instance, while we may not know exactly what wave-function is, we certainly know what wave-function is not—and it is not physical, or local. Here is an abstract, transcendent substance that is indispensable to scientific truth.

For the first time in scientific history, we are confronted empirically with a quasi-spiritual substance. And what do we do? We quickly label it "wave-function" and move on, as if we have discovered simply one more natural/physical phenomenon. The reality of the situation however is quite different. The reality is that science has now discovered that

---

[137] God may act in nature in any number of ways, but among these, *DQA is the only way that is suggested by nature itself!* Consider: according to quantum theory there are no "naturalistic" efficient causes that determine quantum event outcomes. Yet quantum event outcomes nonetheless do obtain determined states! This is an inexplicable (indeed paradoxical) fact of quantum reality and (as we have seen) may be taken to suggest that quantum events are determined by *"non-naturalistic"* efficient causes! That is, by causes *beyond* nature itself! What seems to be required (indeed *necessitated*) by quantum law is a *transcendent* determiner of quantum indeterminacies!

the material world and all the physicality thereof is but a shadow on the wall of a Platonic cave. The illusion here is not reality, but rather "physicality." The eternal substance has finally made itself known and it is incorporeal, indeed spiritual.

Particular Providence and the Particle-pusher Problem

Prior to the quantum revolution there was no good "non-interventionist" way for conservative theologians to account for God's objective actions in nature and the world. As we have seen, particular providence, though seemingly unproblematic on the face of it, is actually fraught with profound scientific and theological difficulties.

At the heart of these difficulties is the scandalous "particle-pusher problem." (The term "particle-pusher" being a derogatory allusion to the classical theological problem of understanding how God, as a Spirit, can possibly interact with particulate matter).

Though seemingly innocuous at first blush, the particle-pusher problem is wickedly knotty. What's more, as a critique, it has teeth, having effectively devastated the Intelligent Design Movement virtually overnight for its failure to postulate a viable mechanism whereby the purported "ID" in question is being imparted.

After a brief season of success in the 1990's, the ID movement rather abruptly fizzled-out, due in large part to critical review and response from both the scientific and theological community. This swift failure owes to (1) the ID movements crude and arbitrary god-of-the-gaps-style supposition, and (2) the ID movements miserable failure to posit (or even suggest) a possible mechanism of design impartation—that is, apart from supernatural interventions that transgress natural laws and contravene causal regularities.

Simply put, one cannot set forth a serious scientific model when that model is based upon episodic miracles. ID theory then, was doomed for failure from the outset, on account of its selective supernatural vision of natural history. Consider, ID theorists claim that naturalistic evolution, while sufficient to account for the vast majority of biological novelty, is nonetheless incapable of accounting for those biological structures deemed "irreducibly complex" (a term both *coined* by ID theorists, and conveniently *conferred* upon natural systems by ID theorists).[138]

The ID scenario of natural history is therefore long periods of slow, gradual, naturalistic evolution punctuated by the miraculous episodes wherein "irreducibly complex" biological systems appear ready-made, in, as it were, a puff-of-smoke.

In the ID scenario, evolution (though more than 99.999% efficient all on its own) would on occasion come up against a biological hurdle so devilishly complex and hopelessly insurmountable (such as, say, a bacteria flagella[139]) that God would be prompted to miraculous action, intervening on the spot, so to keep nature moving smoothly and steadily along.

However, the notion that God would (in Humean terms) "violate" His own divinely decreed laws of nature to enhance the motility of particular species of bacteria failed to convince serious theologians.[140] In the end however, it was the failure to set forth a viable mechanism of design impartation that proved the movement's ultimate undoing.

It didn't take long for serious thinkers (and not-so-serious thinkers) to realize that ID theory was little more than an orchestrated attempt to discover biological structures seemingly beyond the cumulative creative capacity of natural selection.

Ultimately, ID theory is a two-fold assertion, claiming that: (1) biological structures exist within nature that cannot be accounted for on the standard naturalistic evolutionary account; and (2) that these biological structures are a direct product of intelligent design.

---

[138] The term "Irreducible Complexity" was first coined by Michael Behe in his book *Darwin's Black Box: The Biochemical Challenge to Evolution* (New York: Free Press, 1996), wherein an irreducibly complex system is defined as "a single system which is composed of several well-matched, interacting parts that contribute to the basic function, and wherein the removal of any one of the parts causes the system to effectively cease functioning." (Behe 1996). Critics of IC have pointed out that Behe fails to either recognize or appreciate the innovative power of evolutionary "co-option", wherein a biological structure adapted for one purpose, becomes 'co-opted' through natural selection for an altogether different purpose. cf. Kenneth Millers, *Finding Darwin's God: A Scientist's Search for Common Ground between God and Evolution* (New York: Cliff Street Books, 1999).

[139] Or, "Exhibit A" in ID polemics.

[140] It is reasoned that God would no more violate one of His own divinely decreed "natural laws" than He would violate one of His own divinely decreed "moral laws."

It seems that ID theory will never be anything more than a registry of currently unaccounted for biological structures. Filling (temporary) scientific gaps with miraculous events has never been a genuinely persuasive maneuver. It didn't sway Victorian scientists and theologians and it hasn't swayed 21st century scientists and theologians either. Not only is such a maneuver intellectually transparent, it is discredited; and long discredited at that, being but a warmed-over version of William Paley's (1743-1805) long abandoned watchmaker argument.

I simply cannot imagine that God would break naturalistic evolutionary precedent to personally assemble bacterial molecules into flagellar propellers, motors and rotors. The entire "theory" is arbitrary and inelegant. What's more, gap-arguments, insofar as they presuppose divine interventions, are intrinsically vulnerable to particle-pusher problems.

The problem for ID theory is that any *successful* argument for objective divine action will of necessity have to find a way around the "particle-pusher problem", which is *the central paradox of providence!* ID theorist's clearly haven't done this; ergo, their lack of success.

Naturalizing the Divine Hand?

To better appreciate the significance of quantum mechanics for divine action, let us try a simple thought experiment: Imagine we live prior to the quantum revolution and are attempting to fathom the workings of the divine hand. With the mechanics of divine action in mind, we begin an inquiry: What exactly is the means whereby God interacts with the physical substrates of the material world? Does God utilize a field or force of some kind? Does God, as a disembodied agent, temporarily manifest in physical form? Or does God perhaps use some form of telekinesis? And if the latter, what is the telekinetic force akin to? What is its substance? What are its constituents?

How are we to make sense of a disembodied agent acting causally in nature and the world? Unlike physical entities and forces (which are, in principle, detectable) God is a Spirit. Ergo the paradoxical nature of providence and the God/world interface. One option would be to imagine that God utilizes physical force phenomena (such as electric charges and magnetic fields) to manipulate particulate matter into His bidding.

Such a method however, (being physically-based), would reduce the divine hand to that of a competing physical force among nature's already existing physical forces. In this view God would be akin to an unseen 'dynamo' who during instances of divine action simply turns on the juice, so to speak, in order to generate the requisite amount of force necessary to bring about an intended end. Such a notion however, would severely belittle the divine hand by equating it with mundane physical processes. Of course, to imagine the divine influence as being somehow on par with the mundane 'pushes and pulls' of ordinary natural force phenomena is to devalue the theological doctrine of providence by 'naturalizing' what is, in reality, a *divine* action. In doing so we relegate the power of God unto a category of naturalistic process—not unlike a 'fifth force.'

In this, we not only subordinate the divine influence to the level of mundane natural force phenomenon but also render the divine influence susceptible to scientific detection!—at least in principle. After all, if God exerts physical force upon physical objects in the same manner as the natural forces, then this exerted force would be just as detectable, measurable, and identifiable as any of the other natural forces (or forces *within* nature).

This option thus clearly fails. For, God would never allow His discrete actions to be scrutinized by science. As a possible alternative view, one might invoke the "*mene mene tekel upharsin*" option of Daniel 5, and posit that God simply spontaneously materializes, sporadically popping in and out of existence in three-dimensional material form.

The obvious problem with such an 'intermittent materialization scenario' is that it imagines God materializing in physical form and exerting physical, effectual force on particulate matter, thus landing us right back at the particle-pusher problem!

In such a scenario, divine agency in the world is no different than human agency in the world, with God agentially directing particles towards a desired end—not unlike a billiards player. Here, once again however, the divine hand is reduced to the level of mundanity—this time, unto human agency.

In all actuality however, any similarity between divine action and human agency will be less like human bodies moving matter and more like human *minds* "willing" human bodies into action. The question then becomes 'how does the divine Mind influence matter?' And we are right back at the central mystery: how can a God who is Spirit possibly interface with the material constituents of the cosmos? How does immateriality influence and/or effect materiality? Of course, as to how this might occur we can only speculate.

For instance, a person of sufficient imagination might imagine that God (as a disembodied agent) is privy to a uniquely and particularly designed particle unbeknownst to mankind. Say, a particle with both spiritual properties as well as physical properties, thereby allowing God (whom recall is Spirit) to interact with the spiritual side of the particle just as easily as the physical side of the particle interacts with the material substrates of the world.

Incredibly, a scenario similar to this seems the only way out of the divine action dilemma! Indeed, for a disembodied God intent upon acting within the world, the creation of just such a particle would be not only supremely clever, but *logically necessary!* That said, the discovery of just such a particle existing within nature would be highly suggestive: particularly to those who already subscribe unto the notion of a Personal Creator-God.

Here is a good example of how science might (or might not) lend credence to the particular claims of Christian doctrine. That is, by revealing how nature in one way or another panders to the precise ontological necessities of precise theological claims. Here is a place where present-day believers have it over earlier generations of believers. Present-day believers have the opportunity to fathom creation (and creations hidden workings) in greater detail than any previous generation. From the secret workings of nature's hidden laws to the particular properties of nature's particulate zoo. Nature is manifest before us. And within its naturally occurring shadows can be discerned the silhouette of the divine.

We have at our disposal a wealth of knowledge as to the workings of the universe, from the exchange rate of matter and energy, to the relative strengths of the fundamental forces and constants. Science has even penetrated the first second of the creation event!

Knowledge of this sort provides us with a window—a window not simply into the workings of the universe, but into the very Mind of God itself, in its facet as Creator. And from this, our understanding of nature (and our understanding of the *nature of nature*), it might be possible to gain insight into how the Mind that constructed this very world of laws and processes might go about acting within it.

The universe is a direct creation of God, and as such, may provide insight into the nature of the God/world interface. The structure of nature's laws and processes may themselves hint at, or suggest, a manner, means, or mechanism of divine interfaction? So while we do not have direct access to the divine Mind (apart from revealed Scripture) we *do* have direct access to creation, which is a product of the divine Mind. As such, it might be possible to take (1) what we know of God and God's actions from Scripture, and (2) what we know of nature and nature's processes from science, and come up with a likely model of divine interfaction. One that is perhaps suggested by both Scripture *and* science!

It seems to me that any hope of fathoming the otherwise unfathomable God/world interface will most likely rest in a proper understanding of both divine *and natural* revelation.

For instance, (and as mentioned above) taking what we know from particle physics we might imagine that God utilizes some sort of (divine?) force-carrier particle. Nature here provides us with a lead. As we have seen, God is not likely to act in the guise of an invisible force physically pushing and pulling particles around, as such action would be scientifically detectable—not to mention in violation of the laws of energy conservation! The same of course, would hold if God acted through physical manifestation, as a mere Agent among agents.

It thus seems to me that God, as a disembodied agent, most likely simply "wills" the actions of His intentions into being by determining and/or specifying the outcomes of otherwise indeterminate quantum events!

Returning to the above-mentioned notion of a "divine force-carrier particle"; if we tweak our notion just a bit, say, to that of a 'divine-intention transmitter/elicitor particle' that necessarily converts "*divine will*" into "*divine action*", we may be onto something.

This imaginary particle of ours, which is of course purely hypothetical, would have the capacity to convert divine *willful intent* into divinely desired *physical action!* Such a particle would take pure "force of will", and cash it out as pure "intentional action" such that 'physical interfaction' is no longer required! Such a particle therefore would have be equipped to not only *engage physical matter,* but also *be engaged by non-physical spirit!* Let us therefore call this hypothetical particle a "Spiritron."

Such a particle, insofar as it would be intended to serve as an interface between the spiritual and the physical, would need to possess a very particular ontology. Namely, *a dual ontology*, for, to be either matter-like or mind-like. The former for "engaging", the latter for "being engaged."

Now, a question for the reader: Assuming that the Personal Judeo-Christian God designed and created this universe—and did so with the express intent of acting objectively and efficaciously within it—*would we and/or should we expect to discover within this universe the existence of a particle with properties as purposefully particular as those of the hypothetical "spiritron?"*

Having personally pondered this question as unbiasedly as my mind and spirit will allow, I am going to have to say *no! I do not think we either would or should expect to discover such a particle existing in nature. The reason being that the ontological demands of such a particle would simply be too great!*

A particle capable of interfacing with both matter and mind is a pretty tall ontological order—even for a 'Divine Engineer', in that spirit substance and matter substance are logically antithetical substrates! This being the case, the chances of actually discovering a "spiritron" (or 'spiritron-like' particle) existing objectively in nature is next to nothing!

For this reason and others, the actual empirical discovery of a particle with the same antithetical properties as the so-called 'spiritron' would be monumental—especially to those with the scientific and theological acumen to recognize the particle for what it is: *an absolute necessity for a "theistic" universe!*

A theistic universe (*that is, a universe designed and created by the Personal God of Judeo-Christianity*) would be expected to be readily equipped with key features—features resonant with a reality created by a spiritual being intent upon acting within that reality—features that we would have no good reason to expect the universe to have, if it *were not* engineered by the Judeo-Christian Deity!—features like "the spiritron", being both consistent with Christian theology and necessary for Christian reality!

Of course, speculative talk of such a particle is ultimately pointless—unless, that is, an atomic entity fitting the quasi-physical description of our hypothetical "spiritron" actually exists.

Recall the unique nature of the spiritron: half mentalistic and half physicalistic, for receiving from God and delivering unto matter. In such a world (i.e., a world wherein spiritrons actually exist) divine action would not arise as a consequence of God physically and directly manipulating either matter and energy, or physical forces and fields. Rather, divine action would arise via God divinely determining the ontological *indeterminacies* of quantum reality! *That is*, via God instructing, and/or specifying the inherently "unspecified" actions of the constituent particles of quantum reality!

The properties of such a particle would be wholly remarkable, and unlike anything discovered in the world of physics up to that point. The existence of a particle so theistically conspicuous existing in the real world would be profoundly suggestive. Its presence in nature would be immediately apparent; particularly to theistically-minded physicists who could not fail to recognize its theological relevance.

The existence of the so-called spiritron would speak volumes as to the true nature of reality, and provide compelling empirical evidence for their being an overall purpose to reality. And a very particular *'overall purpose'* at that. Namely, that revealed in the Judeo-Christian Scriptures!

As the reader has by now undoubtedly figured out, the spiritron *does in fact exist*; and is of course, "the quantum." The theistic utility implicit within this unique, dual-natured entity continues to amaze me. The sheer elegance and ingenuity of its design is a source of on-going awe and inspiration—*engineered by God himself, to ontologically oblige and accommodate the metaphysical goings-on of theism within the material world.*

Here is a species of design unlike anything we've ever seen, or imagined.[141]

Here is theistic utility and theological purpose set within the ontological foundations of physical reality itself.

What we ultimately have then, is a physical universe grounded upon a layer of metaphysical process. And the latter we call quantum reality.

---

[141] Being wholly unlike the innumerable gap-arguments from biological complexity, which have been the standard M/O in natural theology for the past three centuries.

The million-dollar question for the atheist then, is this: why should the universe appear so readily prepared to pander to the particular precepts of Christian theism, if there is in fact no God?

It seems to me that the simplest, most obvious, most logical, *and hence best*, answer to this question is that the universe *appears* readily prepared to pander to the particular precepts of Christian theism, because the universe *simply and genuinely is* readily prepared to pander to the particular precepts of Christian theism.' Thus, God exists. And not just any God—*the God of Judeo-Christian theism*.

Here then, is a case where *appearance is reality!*[142]

Mind and Observership

As we have seen, a theistic universe necessitates a dualist ontology. And as we have also seen, quantum mechanics affords just such an ontology. At the heart of this quantum ontic is an entirely new order of existence—namely, *potential existence,* and it is neither physical nor spatiotemporal, but rather *abstract* and *informational.*

Just as Einstein demonstrated that mass can be converted to energy (and vice versa), so Max Born (a close friend of Einstein) demonstrated that particulate matter and/or energy can be converted into abstract potentiality waves (and vice versa), or what Einstein referred to as "*Gespensterfelder*"—i.e., "ghost waves."[143]

Einstein's ghost waves (which are of course quantum wave-functions) are about as close as any essence or substrate can get to being a spirit-like substance. The most amazing aspect of all this is that these spiritualistic ghost waves play a central role in the realization (or "actuation") of physical reality. What's more, the effects of these overtly spiritual waves can be detected *scientifically!*—due to the spiritual/physical dualism of quantum reality.

As Einstein demonstrated mass/energy correspondence, so quantum theorists (e.g., Max Born, Werner Heisenberg, John von Neurmann,

---

[142] Recall that the connection between quantum physics and Christian doctrine (e.g., free will, divine action etc.,) was recognized almost immediately (and independently) by a number of quantum pioneers, such as Arthur Holly Compton, William G. Pollard, Karl Heim, Eric Mascall, Frederik Belinfante, et al.

[143] Louisa Gilder, *The Age of Entanglement: When Quantum Physics was Reborn*, (New York: Random House, 2009).

Eugene Wigner, et al.) demonstrated that matter can exist as pure mathematical potentiality, such that the entirety of this physical universe ultimately reduces to pure abstraction.

This has been widely known and accepted since the early 1930's when the Hungarian mathematician and polymath John von Neumann demonstrated that quantum mechanics literally reduces to a set of abstract mathematical states.[144] (It is interesting to note that von Neumann [along with another Hungarian genius of the period Eugen Wigner[145]] believed that consciousness could actually interact with and/or effect wave-function).

Consider: in ordinary scientific theories mathematics is used as a mere tool to describe and/or explain a given object, process, or phenomenon. In quantum physics however, the line of demarcation between abstract mathematical description and concrete reality blurs irrevocably. The reason being, that the mathematics employed to describe quantum reality, becomes the reality itself!

There is a metaphysical threshold here that needs to be recognized and acknowledged. Quantum reality is the existential edge of the physical universe. And there is just enough overlap between our physical world and that ghostly realm that lies just beneath that we can learn something significant about the ultimate nature of reality.

Consider, at the quantum level, there is a blurring between physical objects and their abstract mathematical structure. Physicality it seems, turns to pure mental/abstract construct, such that the realm beyond us is not only incorporeal but also noumenal! What's more, this is an active noumenality, for, it takes physical particles and cashes them out in terms of abstract mathematical potentiality.

How very peculiar. How odd that we should discover *not only* that there is a realm beyond, but that this self-same "realm" cashes out present physical states of reality in terms of future mathematical possibilities! The picture that emerges in one's mind is that of a material world steadily percolating into existence via the seemingly willful determinations of an active mathematical construct!

---

[144] John von Neumann, *Mathematical Foundations of Quantum Mechanics*, 1932, translated by Beyer, (Princeton University Press, 1996)

[145] Wigner states: "It was not possible to formulate the laws (of quantum theory) in a fully consistent way without reference to consciousness." See Wigner, *Symmetries and Reflections: Scientific Essays*, (Indiana University Press, 1967).

However, if the constituents of quantum physics take on the character of objectively existent mathematical constructs, the question begs: "Whose mental constructs are they?" Who is imagining (and thus realizing) these quantum mathematical processes into being? Given the overall evidence from quantum reality the answer is clear.

Consider what quantum physics is actually saying. It is saying that there exists an underwriting mind beneath physical reality! That physical reality literally reduces to mathematical cogitation. And this cogitation results in the ongoing existence of the material universe!

Physical reality is thus ultimately grounded in noetic activity! So while the mathematical formalisms that underwrite quantum theory serve as the onto-existential foundations of physical reality, the mathematical formalisms themselves seem to be grounded in a form of divine noetics. A wondrous picture emerges: that of God cogitating the universe into being, construct-by-construct, determination-by-determination!

As we have seen, the bedrock of quantum reality is wave-function. Wave-functions are purely mathematical and entirely abstract—and by "mathematical" here, I mean statistical! Ergo, the above claim that quantum reality and in fact all reality ultimately reduces to a mathematical sea of abstract potentiality.

At the same time, quantum reality is slightly more than just 'abstract potentia', as there are subtle mind-like aspects implicit within wave-function phenomenology. In fact, *so* distinctively mind-like is the quantum substrate that (as just alluded to above) as early as 1932 John von Neumann posited that consciousness can efficaciously commune with wave-function for, to induce collapse! This view was further advanced by Fritz London and Edmond Bauer in 1939[146], as well as by (the also aforementioned) Eugene Wigner.[147] (Consciousness-induced-collapse models abide to this day in the work and ideas of N. D. Mermin and others).

The problem with such models is that they posit the existence of *macro-level* wave-functions, such that the large-scale objects of ordinary everyday reality (including for instance, *the moon!* [cf. Mermin[148]]) would need to be directly collapsed into existence *via* a conscious observer!

---

[146] See for instance, London and Bauer, 1939, translated by Wheeler and Zurek, *Quantum Theory and Measurement,* (Princeton University Press, 1983).
[147] Wigner, *Symmetries and Reflections: Scientific Essays,* (Indiana University Press, 1967).
[148] See N. David Mermin, "Is the moon there when nobody looks? Reality and the quantum theory", in *Physics Today,* April 1985.

For this reason and others, I am suspect of all such proposals. After all, we only have evidence of quantum process at the quantum level! What's more, "observership" on this view leads to a most ghastly paradox; an infinite regression of solipsistic sets, where each observer bespeaks an antecedent observer, necessarily, and so on, *ad infinitum.*

Such would make for a truly bizarre reality. One where moons, planets, and even people cease to be for mere lack of observation. Fortunately however, this is not the case. For there is only evidence for *micro-constituent* wave-functions. Even so, some have gotten creative and supposed that the logical terminus of the above-mentioned "observational regress" is God, who is of course the ultimate observer. And I say, this would indeed be the case *"if"* macro-level wave-functions were in fact a ubiquitous part of ordinary, everyday reality. However they are not. And so it is not the case. Again, there is only evidence for micro-constituent wave-functions.

Let us therefore leave God's mysterious 'quantum activities' where we found them: *at the quantum level! For, as we have seen, there are very good reasons why quantum reality is as it is. And these are directly linked to Christian theism/theology.*

In our quest for divine insight then, we must take reality as we find it, and not as we wish it to be. It is only when we are open to truth, that we are able to find it.

And what we have found so far, is that while there is no place for quantum collapse at the macro-level (and hence no need of God as the logical terminus of conscious observation), there are quantum wave-functions at the micro-level that in dire need of a transcendent consciousness to determine the indeterminate. As I have said all along, quantum reality, while not directly accessible to human consciousness[149] is readily suited for direct access by *divine Consciousness.*

---

[149] If wave-functions are in fact preternatural/mind-like entities, then it may be that only *other* "preternatural/mind-like entities" (such as say, human souls) can engage them. Such a postulate would go a long way towards accounting for mind-brain interfaction—as key cognitive brain functions involve quantum collapse processes (and hence room for transcendent determinations). If we are indeed tripartite beings as Scripture informs us (1 Thess 5:23), then, human souls (or spirits?) may be able to do what "consciousness" cannot: that is, directly interface with those wave-functions of the brains neurotransmitters, for, to "puppeteer" the body, so to speak. This would of course take place via the soul (or spirit?) determining key indeterminacies within the brain. What's more, such a mechanism would be ideal for God to both revelate and inspire Scripture (not to mention induce the dreams and visions of the prophets). Understand that neuroscience is still in its infancy, and while

Particular issues aside, one thing is for certain; there is an elusive mind-like element to quantum reality that we have yet to pin down; one that von Neumann, Wigner, and a handful of others were perceptive enough to pick-up on. I am forever sympathetic toward particular aspects of these early mind/wave-function-interaction models. After all, I do contend that the divine Mind interfaces with wave-function—albeit, to *determine;* not to collapse.

In my view then, the only mind capable of interfacing with wave-function is *divine* Mind (with the possible exception of human souls and/or spirits—at some unconscious level of course). Subsequently, my view is perfectly consistent with the conservative, mainstream, orthodox view of quantum mechanics which states that human minds cannot induce wave-function collapse, but physical systems of an appropriate size and/or complexity can. (Generally, classical systems).

As for my claims of God, souls and spirits, science is a purely naturalistic enterprise and so doesn't concern itself with the actions of such entities and/or beings. Theologians, philosophers, and metaphysicians however, *do*, as they are not bound by the arbitrary and limiting parameters of science. That said, my theological assertions of God determining quantum indeterminacies are just that: *"theological* assertions." These assertions however, while beyond the scope of orthodox science, are in no way contrary to the *claims of* orthodox science.

I believe that *'where science ends, theology begins.'* And my contentions are a clear expression of this fact.

Quantum Metaphysics

When it comes to quantum orthodoxy, one thing everybody agrees on is that when an observed quantum particle is in an *unobserved* state, it literally ceases from physical being. Unobserved, the particle is a non-localized wave of potentia, and all it takes is a single observation to once again charge the ghostly entity with particulate physicality.

---

inroads have been made in the past few decades, we are still very much in the dark on the question of consciousness. If David Chalmers and others are correct, and mind is truly 'a fundamental element of reality', then a scenario not unlike the one just given, might actually be correct.

What exactly takes place between observations nobody knows. And what exactly this ghostly quantum essence is, we cannot tell.

The reality of the matter is that between physical manifestations quanta appear to revert to a soulish or mind-like essence, not unlike the *nous*, or *soul* substance of neo-Platonism, existing within realms eternal.

Quantum physics has introduced us to a reality long known to exist, but long thought to be beyond the reach of empirical science. Quantum physics has provided mankind with tangible proof of the *intangible!* This is without question the most incredible discovery in the history of science. Nothing is comparable in measure. For we now have empirical evidence that a mentalistic, spiritualistic, soulish essence in fact exists—*and exists objectively*, wholly apart from either opinion, private revelation, or subjective religious testimony.

What we have on our hands is a metaphysical substrate unlike anything in physical reality; apart that is, from mind and consciousness. This substrate, while possessing many uniquely mind-*like* properties (such as being able to communicate *non-locally* over arbitrarily large distances), is not itself *actual mind*.[150]

Rather, wave-function is an essence all its own, with distinct properties all its own—in addition to its various mind-like properties. As such, it's more of an overlapping and/or underlapping essence uniquely set to serve as a medium between mind and matter. The mere existence of such a substrate is highly suggestive of dualism, and can potentially revolutionize cognitive science.

Diehard physicalists however, reject consciousness as evidencing dualism, choosing rather to believe that mental states and consciousness (including their own) is but an incidental by-product of random physiological processes in the brain. But if this is so, why would any

---

[150] Another example of quanta displaying mind-like behavior is the seeming subjective nature of actuation that occurs during controlled laboratory measurements. Here, quanta conform to whatever the experimentalist asks for, manifesting as either waves or particles depending upon the nature of the experimental set-up. If the experimental set-up asks for a wave measurement, quantum reality will accommodate; if the experimental set-up asks for a particle measurement, quantum reality will oblige. Hence, the adage; 'ask a quantum question, get a quantum answer.' So while the quantum substrate exists objectively, it is happy to conform to the subjective inquiries of the experimentalist. Observed quantum properties then will owe just as much to the nature of the mode of the experimental set-up as to the nature of the quantum entity itself. It thus appears that the line of demarcation between subject and object has been irrevocably blurred. There is an ontological grey area that makes it impossible to determine where consciousness ends and reality begins.

physicalist have any confidence whatsoever in his or her own physicalist theory of mind? It would be foolish to put stock in what are purely random mental states. All the more, to believe that our mental states are purely random! Such is the tortured logic of the reductive physicalist.

Isn't it much saner to believe that our waking mental states are deliberately our own, and that we feel *as if* we have free will because we *actually do* have free will (and hence a soul)? It would seem so. Of course, one could counter that just because one "feels like something" doesn't make it so. And in this they would be correct. However, when you consider the fact that we inherently and self-evidently feel as if we have free will (as well as a soul, in that we feel we are distinct from our bodies); alongside the fact that we have now discovered the existence of a soulish essence; and together with the fact that our neural processes reduce to quantum processes involving this self-same soulish essence; I think the reality of the matter is clear: physicalism is false and we are constituted just as Christian anthropology states!

Another clear reality is quantum dualism. That an abstract quantum substance exists is demonstrable scientific fact; pure and simple. There's no two ways about it. There is simply is no good way of explaining away the dualistic, spiritualistic, *theistic* character of reality!

Understand, this is unprecedented news for Christian theism. Quantum physics has cast a silver bullet—and set it deep in the heart of "physicalism." It is but a single bullet, but it is a lethal bullet. Material reductionism is dead in the water, and physicalism is soon to follow. For, if quantum physics has shown anything, it has shown that "physical substance" is definitely not the only substance that exists!

This "silver bullet" is quantum dualism. And bringing an end to physicalism is but one of its theistic benefits. For, quantum process, by its very nature, has rendered reality open to a theistic interpretation.

Mind-Stuff

The ethereal essence unto which this physical universe reduces has been likened to "mind-stuff" by both physicist and philosopher alike. And for good reason. Wave-function is uncannily analogous to thought. For, along with being abstract in essence (and hence non-physical) wave-function is also non-spatiotemporal, existing beyond the three-dimensional confines of this physical universe.[151] Even so, wave-function is objectively real—just not "physically." Wave-function is however objectively real "metaphysically."[152]

---

[151] Wave-functions however, do evolve temporally and spatially within the abstract mathematical domain of Hilbert space, which is a Platonic domain that is over, above, and outside of this physical realm. As for wave-function spatiotemporality itself, it obtains via the evolution of the Schrödinger wave equation, which models the objective phenomenological processes of wave-function! This is an enigmatic state of affairs to be sure. For, not only do wave-functions evolve *outside of* this physical universe (in abstract Hilbert space), but they also seemingly (and simultaneously) evolve *within* this physical universe in that their collapse brings about physically observable effects within the domain of their evolution subsequent to their previous collapse. The only way to make sense of this situation is to assume that quantum wave-function is an ethereal, platonic-like substance that is mathematically representable. Only a metaphysical substance could do what quantum wave-functions do. That is, evolve "non-physically" both beyond (and possibly within) this physical universe. Here is a substance that seems to be both transcendent and immanent simultaneously! Clearly, we are dealing with a metaphysical substance.

[152] Since the inception of quantum theory physicists have debated the question of how best to interpret the quantum wave-function. Questions such as: Does the wave-function represent an objectively existing reality? Is it real? And if so, in what way is it real? Or, is it merely a probability distribution representing limited and subjective human knowledge? At the end of the day, all such deliberations leave us with one question: Is the wave-function real, or not real? I, of course, believe the former to be the case. I believe that wave-functions are genuinely real entities. And we now have very good evidence that this is in fact the case. Wave-functions exist, and they are real! (For an overview of wave-function ontology see Ney, and Albert, *The Wave Function: Essays on the Metaphysics of Quantum Mechanics*, [New York; Oxford University Press] 2013. See also Colbeck and Renner, "Is a System's Wave Function in One-to-one Correspondence with Its Elements of Reality?" *Phys. Rev. Lett.* 108, 150402 [2012] APS). That wave-functions are genuinely real, genuinely existent entities is revealed in the fact that there exists a "one-to-one correspondence" between wave-function and reality! (Colbeck and Renner, 2012). This strongly suggests that wave-functions *are* the reality they represent! The "one-to-one correspondence" between quantum wave-function and quantum reality (for instance, as argued by Colbeck and Renner) speaks both *to* the objective existence of wave-functions and *against* the subjective knowledge interpretation of wave-functions. While the former sees wave-functions as genuinely existent entities, the latter sees wave-functions as mere mathematical tools for generating probability calculations, and having nothing to do with objectively existing quantum waves. Furthermore, the latter subjectivist's see quantum probability calculations as offering epistemologically limited data. Unto these, quantum theory is incomplete (however, they hope for better calculating techniques in the future). Objectivists on the other hand (i.e., wave-function realists) believe that wave-functions provide both an accurate *and complete* description of reality. Unto these quantum theory is a complete theory. And I believe the evidence bears this out, such that quantum mechanics is complete and wave-function realism is true. There is however a catch with regards to wave-function ontology, in that while wave-functions are in fact real, they are also *non-physical*. Here is where the later ontological views of Bohr and Heisenberg are most discerning. Bohr and Heisenberg not only held that quantum theory was complete (contrary to Einstein's epistemological view), they also rejected de Broglie and Schrödinger's assertion that wave-functions are physical matter-waves. We are thus left with a dilemma: for, if

As to the question of what this substance is *in itself,* we have not a clue. It is however enough for me personally, simply to know that spirit-like substances do in fact exist—and exist "empirically."

How incredible is this? Pause for a moment and truly take-in what quantum physics is telling us: *mankind has discovered the existence of a substrate that is not only not physicalistic, but actually spiritualistic!* This discovery is of profound significance for Christian theism—not least, because quantum process seemingly panders to the onto-existential requisites of Christian doctrine!

A postmodernist however, might argue that Western Christianity has no monopoly on quantum metaphysics, pointing out that New-Age spiritualism and Eastern mysticism also claim quantum support. The problem here, however, is that every such "cult" and "ism" invariably invokes a crackpot version of quantum theory!

New-Age mystics and Eastern gurus grossly mischaracterize quantum physics as a mind-over-matter free-for-all, wherein a willing subject can simply visualize (or "observe") the reality of his or her own choosing into being! This of course, is nothing more than New-Age fluff, having nothing at all to do with actual quantum physics. Conversely, when it comes to quantum mechanics and Christian theism, there is wresting of neither physics nor theology such that both retain their orthodoxy. And this is what makes the quantum/Christian connection so compelling.

---

(1) quantum mechanics is complete; and (2) quantum wave-functions are not physical; then I ask, *what* in heaven's name is the ontic standing of wave-function? We can only conclude that wave-functions, while objectively real, are nonetheless non-physical! This brings us to the core of the issue: namely; just what sort of reality *is* wave-function? This much we know: firstly, we know that wave-functions are objectively real because they produce double-slit interference patterns; secondly, we know that wave-functions are non-physical because they have no physical properties; and thirdly, we know that wave-functions are yet incorporeally existent because they are capable of instantaneous non-local influencing over arbitrarily large distances. That said, if quantum physics has taught us anything, it is that *physical* reality isn't the *only* "reality" in town. Wave-function has been likened to "mind-stuff" by both physicist and philosopher alike. Both for its instantaneous communicability and its seemingly willful selecting-out from among potentia. Not to mention the property-status-subjectivity intrinsic to quantum measurement. What's more, wave-functions are ontologically abstract and fundamentally mathematical, such as to be essentially platonistic. Ergo, the transcendent, eternal, nonspatiotemporal character of wave-function. Summing up, whatever class of reality we have here stumbled upon, it is ontologically abstract, structurally wave-like, elementally mind-like, essentially volitional (and so seemingly personal), *and thus quite possibly spiritual.*

There is an inexplicable congruence between the natural ontology of quantum physics and the necessary ontology of Christian theism, such that the former underwrites and realizes the latter, while the latter provides ultimate meaning and expression unto the former. There is a sublime unity in this give-and-take economy such that it is difficult to distinguish where either ends and the other begins. The subtlety and ingenuity of this ontological complement speaks to both the wisdom and savvy of the Creator. For we truly live in a cosmos that is both cleverly and compellingly theistic.

Such are the ways of the Judeo-Christian Creator.

Spirit

We are now well into the 21st century. It has been over one hundred years since Max Planck first formulated quantum theory (1900). Looking back from the quantum revolution unto today, it is tellingly clear that modern science has utterly failed to deal with the fact that quantum mechanics has revealed the existence of a realm that is wholly "other!"

Quantum reality is quite literally a "notional" reality. And yet, paradoxically it is the source of all physicality. Again, this is an empirical scientific fact! One that just so happens to also be a *theological assertion!*

So never again let the *materialist* say that the Christian Faith is groundless. For we now know that the ground of matter itself is spirit!

And never again let the *empiricist* say that the Christian God (who is Spirit) is unscientific because things of the spirit do not lend themselves to empirical investigation. For we now have empirical evidence of a transcendent spirit-substance!

Neither let the *naturalist* say that miracles are unscientific; as they are common-place at the quantum level. And not just in the sense of events taking place in the face of insurmountable statistical odds. But in the sense of events taking place that defy macro-physical law!

Nor ever again allow the *physicalist* to say that man is nothing more than a sum of physical parts. Or that individuals are but fortuitously-evolved meat-machines possessing nothing even remotely like a soul that survives death. For we now know that cognition and/or "mind-stuff" is an objectively existent medium that seems to be not only *ontologically fundamental*, but also existentially elemental being basic to the whole of physical reality!

And finally, never again let the *atheist* say that there is no evidence for the Personal-Creator-God of Judeo-Christianity, for not only has the universes fundamental constants been finely tuned for life[153], but more tellingly, the universes ontological structure appears to have been engineered to sustain the self-same spiritual processes requisite to a Christian realist worldview!

In turn, let the *dualist* now confidently state that for the first time in history there is empirical proof of both a corporeal substance and an incorporeal substance.

And with the last word of them all, let the *Christian* now confidently declare that science, rather than controverting Scripture, actually corroborates Scripture, confirming the ontic grounds of doctrinal truths long avowed by the Christian community.

These facts, along with others mentioned, have been grossly underreported by the materialistic scientific paradigm of present. Even so, the facts remain; contrary to three-hundred years of scientific materialist dogma, *mind precedes matter*, and not the other way around.

I ask then, why should we continue to believe the now scientifically discredited modernist supposition that physical reality is the whole of reality? Or the physicalist atheist assertion that the death of the physical body equals the death of the immaterial person? *We shouldn't!* For, the fact of the matter is that physicalistic materialism is now dead. And it is dead at the hands of empirical science! Understand, this is not an opinion. It's a fact. *A fact of quantum physics.*

That mind precedes matter should come as no surprise to the Christian. What is God, after all, but a disembodied Mind, Spirit, or Soul. For science then, to now inform us that matter (and in fact all physicality) derives of a transcendent mind-like incorporeality is a confirmation of the Scriptural narrative far greater than any believer could have possibly hoped for.

---

[153] Barrow and Tipler, *The Anthropic Cosmological Principle* (Oxford: Oxford University Press, 1986).

Chapter Six

Quantum Uncertainty and the Subtlety of God

> What can be seen on earth indicates neither the total absence, nor the manifest presence of divinity, but the presence of a hidden God. Everything bears this stamp.
>
> —Blaise Pascal, *Pensées*

As it happens, a quantum ontology is necessary for a theistic economy. And the principal currency is wave-function. This is not at all surprising given the disembodied nature of the Judeo-Christian God. What is wave-function after all but an incorporeality closely akin to thought? A dynamic mental construct from whence all future states of the material world come into being. First as potentiality, then as actuality, each wave-function collapsing away to become a physical constituent of the material universe.[154]

---

[154] In speaking of Platonic realism throughout this work, I have intentionally ignored issues of nominalism (cf. Fictionalism, Neo-Meinongianism, etc.) as I believe the empirical fact of wave-function realism settles the question once and for all. As regards realism and Christian theism however, there is an apparent conflict between the supposed necessary existence of abstract objects and the sovereignty-aseity standing of God. Consider; according to the sovereignty-aseity doctrine God exists independently and self-sufficiently, in and of Himself. He depends upon nothing or nobody for existence. By the same token, all that is not God is creation, and is therefore dependent upon God for its existence. Again, God is dependent upon nothing. Everything else, is dependent upon God. The basic problem

Each actuation is couched in indeterminism thus shrouding the divine hand in a cloak of ambiguity. Here is a place (among other places) where the fundamental structure of quantum physics so elegantly panders to theism. For, behind the indecision of every underdetermined quantum event, lay the full complement of God, waiting to will such events to their completion. God, in acting from beneath the quantum fog, retains both His subtlety and ineffability. There is a fine line here between God as acting agent *in* the world, and God as Creator and sustainer *of* the world. For, both are realized via the grist of quantum actuation.

Here we glimpse the mode and mechanism of *creation continua*, as well as the means whereby God can (if He so chooses) manipulate every individual particle in the cosmos. Here is a scientifically confirmed mechanism whereby a disembodied transcendent Deity might control the entirety of physical reality from the bottom-up without violating a single law of science and without ever once giving either Himself or His actions away!

---

then, is that certain abstracta (i.e., various propositions, properties, and objects) seem to also exist necessarily, and therefore independently of God! The resulting consequence is an apparent infringement upon the sovereignty-aseity of God. All is not lost for theology however, as a number of resolutions have been set-forth. The main popular being: Absolute Creationism, Universal Possibilism, and Divine Conceptualism. Out of these, I find divine conceptualism to be the most coherent. In addition, it is strikingly in-step with quantum physics! After all, has not contemporary quantum physics shown that certain abstract objects exist objectively while being mentalistic in character? Does this not fit ideally with the aseity-sovereignty-friendly doctrine of divine conceptualism? Certainly it does. These particular issues are nothing new to Christian theology, Augustine was a conceptualist, as was Abelard, Boethius, William of Occam, Descartes, Locke, Leibniz, et al. In adopting some or other form of conceptualism then, we find ourselves in excellent philosophical, and theological company. On a different note, a conceptualist understanding of wave-function could actually shed light on a mystery pertaining the origin of the universe. Consider, atheists often contend that the Creator-God of Judeo-Christianity is no longer necessary to account for the origin of the universe, claiming instead that the universe might have simply popped into existence via a vacuum fluctuation. There is a major problem with this idea however; for, if, as it is argued, space and time came into existence *at* the Big Bang, (along with the laws of physics; *including quantum physics*), then there could have been no quantum potentiality wave in existence *prior to* the Big Bang, for, to give rise unto it! For, not only was there no "before" the Big Bang (as time and space began with it), but neither was there quantum processes, as these too only came to exist at the Big Bang! Conversely, on a quantum conceptualist account, things gel quite nicely. For, if, as I contend, quantum wave-functions are (like other abstract objects) somehow conceptually grounded in the divine Mind, then the potentiality wave (which, recall, I likened to divine middle-knowledge) could have existed in the Mind of the Creator prior to initiating the act of creation!

Thomas Tracy states,

> In an indeterministic world of the right sort, it would be possible for God to act through the structures of nature, yet leave those structures entirely undisturbed. God's action would realize one of the alternative possibilities generated within, but left open by, the causal history of the world. This would alter the direction of the world's development so that events evolve differently from how they would have had God not so acted; but it would do so without displacing natural causes. The result is a non-interventionist and non-miraculous particular divine action.[155]

This is of course, the exact sort of world we find ourselves living in! And it is indeed "an indeterministic world of the right sort", having been perfectly engineered for to allow the Judeo-Christian God to act clandestinely; preserving the regularity of natural law, and in turn the integrity of human free will. The cosmos may be 'fine-tuned' for life, but it's *engineered* for a Judeo-Christian economy!

Ontological indeterminism of this quantum sort is a stunning feature for the cosmos to have. Not only because it's logically counterintuitive, being thoroughly contrary to the guileless rationality of the Newtonian universe (where causes always precede effects, and effects always have sufficient determinants), but because it's so blatantly metaphysical!

Is it not uniquely absurd that the naturalistic cosmos we perceive at the macro level is ontologically grounded in a hidden foundation of pure metaphysics! Certainly. But why such an odd and seemingly arbitrary brand of dualism? Well, as it turns out, what appears to be arbitrary is in reality necessary. And not only with respect to physical process. For, as we have seen, quantum process panders quite distinctly to a Christian realist economy. And in so doing, divulges the identity of the creator as that of the Judeo-Christian God!

Quantum physics has brought much to theism, both in terms of resolution and explanation. One need simply consider to aforementioned "particle-pusher problem." This long-standing problem in both classical and modern theology was resolved virtually overnight via the

---

[155] Thomas F. Tracy, "Theologies of Divine Action";*The Oxford Handbook Religion and Science,* ed. Philip Clayton and Zachary Simpson (Oxford: Oxford University Press, 2006).

implementation of quantum indeterminacy.[156] Likewise, with respect to the issue of freewill within a deterministic ontology.[157]

Many other examples could be given. Even as many as to compel an informed believer to state (after the spirit of Wigner) that the 'unreasonable effectiveness of *quantum indeterminacy* for the realization of Christian theism is a wonderful gift which we neither understand nor deserve.'

I cannot overstate how incredibly well suited quantum physics is for the Christian faith. A believer in Newtonian times would never have dreamt that such an 'about-face' with respect to ontology was even possible. And yet it is *our reality*. The great irony is that this unimaginable truth is lost on the vast majority of believers.

What is this unimaginable truth? In a nutshell, it is that the physical world is grounded in what appears to be the constructs of a disembodied Mind. And what's more, the phenomenology of this abstract substratum panders quite ingeniously to the metaphysical requisites of Christian doctrine—while also resolving a host of seemingly intractable theological paradoxes. One need only think of the ages old theological paradox of divine Sovereignty vs. freewill. Only in a quantum world can both be wholly true!

You see, that which is random from inside the universe, may be fully determined by God from outside the universe. The rules of quantum physics allow for just this sort of behind-the-scenes action.

Let the atheist then scoff at the various metaphysical accounts of Scripture. But also let the atheist know that for every preternaturalism of Scripture there is a corresponding quantum preternaturalism; both to (1) ontologically and phenomenologically ground the former, and

---

[156] The implications of quantum indeterminacy for divine action (and the particle-pusher-problem) was recognized early-on; even within decades of the formulation of quantum theory. In fact, physicist-theologians William Pollard and Karl Heim wrote on the subject as early as the 50's. Alas, their insights fell largely upon deaf theological ears. Subsequent physicist-theologians such as Ian Barbour and Robert John Russell however, have continued in the ideas of Pollard and Heim before them, looking to quantum physics for insights into divine action and/or around particle pushing! As Barbour tersely explains: "God does not have to intervene as a physical force pushing electrons around but instead actualizes one of the many potentialities already present." (Barbour, *Nature, Human Nature, and God*, [Minneapolis; Fortress Press, 2002]). The ideas of Pollard and Heim thus live on.

[157] Arthur Holly Compton championed the idea of freewill *via* indeterminism as early as 1931. (Science, vol.74, Issue 1911, August 14, 1931).

to (2) demonstrate that present-day scientific reality is every bit as metaphysical as Biblical reality!

As for the unbelieving pseudo-rationalist, let him appeal no more to the "absurdity of miracles" as an excuse for not believing. For, as aforesaid, miracles are commonplace at the quantum level!

The Majesty of Being

As we have seen, physical reality emerges from a roiling tumult of sub-atomic indeterminacies.

Yet, from this uncertain sea there collectively emerges the higher order intelligibilities, laws and regularities of the large-scale material world! This is a seemingly miraculous fact of reality. Yet, again, it is our reality. From uncertainty comes certainty. From chaos comes order, symmetry, beauty. All that's necessary for a rationally intelligible world.

Who could have possibly imagined that the structural symmetries of large-scale reality actually draw their stability from the abyss of uncertainties that is quantum reality? After all, is it not infinitely more reasonable to suppose that "from chaos" would come only "more chaos?" It would seem. But again, from quantum process, the unimaginable quite commonly occurs.

Still, given the nature of quantum reality one would expect the world to be a random sea of fortuitously interacting particles. Instead, these indeterminacies, being stochastically constrained, work together so as to give rise (writ-large) to a stable macro-realm where lawful regularities allow for the development and practice of science.[158]

From stochastically constrained randomnity then, there emerges a reality shot through with rational intelligibility. We subsequently discover a world of profound structure, sublime beauty, and nested complexity. From galaxy clusters, to galaxies, to planetary systems, to

---

[158] Consider Fermi-Dirac, and Bose-Einstein statistics. These, while given to indeterminism at the quantum level, serve writ-large to ground determinism at the classical level! Robert Russell writes, "How strange is it that the classical, everyday world, where Boltzmannian statistics point to causal determinism, is actually the product of a quantum world whose FD and BE statistics points instead to ontological indeterminism!" (Robert John Russell, *Cosmology: From Alpha to Omega; The Creative Mutual Interaction of Theology and Science* [Minneapolis: Fortress Press, 2008]). How poetically ironic that the self-same deterministic regularities that materialists use to disregard God and providence, ultimately derive of micro-level indeterminacies that are both grounded in, and determined by, the Divine Mind.

eco-systems, to life, to sentient life, to human life itself, endowed as it were with a God-consciousness; an inherent thirst after the transcendent, the eternal, the divine. Not to mention a personal conscience whereby we can come to know the true self within, as well as the Divine Self without. (This latter of course being God).

But why should we expect randomnities, even if stochastically constrained, to lead ultimately to persons capable of pondering the majesty of the universe? We shouldn't. That's the point. Not unless there's a Grand Mastermind at work both beneath, above, behind, and throughout the randomnity. Which is exactly what I posit!

How else do we account for the fact that a finite number of particles (10^{80} baryons in the visible universe[159]) acting over a finite period of time (13.7 billion years) managed to assemble themselves into ever-increasingly complex aggregates of matter, until that matter itself attained sentience? And not just *any sentience*—but *sentience* marked by (1) a robust God-consciousness, (2) a deep-seated awareness of self, morality, and existential finitude, and (3) a profound thirst for spiritual wholeness?

The uniquely human quest for wholeness via the transcendent (albeit personal) "otherness" that is God has been the defining mark of our species ever since modern man appeared on the scene some 40,000–60,000 years or so ago.

When modern man appeared, he appeared abruptly. And when he appeared, he appeared with religion in hand—or rather, in heart, mind and soul. From the sky-god cults of early hunter-gatherer clans to the matured monotheism of the Yahwistic religions, man has always known something of a beyond. For, knowledge of God is both within us, and before us. And the more we modern humans learn, the clearer our picture of Deity becomes. Our generation alone has discovered areas where science and theology appear to overlap in profound and wonderful ways. From the anthropically fine-tuned constants of nature to the theology-pandering processes of quantum mechanics. The *exterior world of man* is now beginning to reflect of the *interior world of man*. And the reflection is undeniably theistic.

The reality that arises from quantum randomnity is one of supreme structure, order, and beauty. It is deeply rational and highly intelligible.

---

[159] Penrose, Roger, *The Emperor's New Mind: Concerning Computers, Minds, and the Laws of Physics* (Oxford: Oxford University Press, 1989).

Its mathematical elegance and economy alone begs for theological return. Even quite apart from the distinct theological purposes that are "one" with the universes ontology.

That such a marvelously purposive order arises from utter randomnity at the quantum level begs a number of serious questions; such as: How is it that individual quantum randomnities know to conspire for to manifest large-scale coherence? What's more, why this particular reality? Why this uniquely tailored network of bio-friendly laws? And why is it that the highest order of complexity in the universe happens to be God-conscientious beings with moral and ethical centers? Is this a clue as to "why anything at all?" A glimpse into why this "something" would be better than "nothing at all?"

These are fascinating questions, born of fascinating facts. The most fascinating of all being that the universe appears to have been painstakingly prepared for our existence! And "prepared" as it were via physical law—the same overarching laws alluded to above. The very same laws that owe to quantum randomnity!

These upper-level, large-scale laws are emergent, arising from the lower-level stochastic laws of quantum chance. As for the quantum probabilities themselves (and their associated indeterminacies); these arise from a platonic ethereality! At least, according to physics orthodoxy.

The cosmos appears to have been engineered (and/or "fine-tuned") for the evolution of life[160]; and human life in particular.[161] It is my contention that this cosmic foresight extends to human spirituality as well, with quantum law pandering as it does to Christian theism.[162]

God is subtle. God is understating. But God is there. And for those with eyes to see, so are the pointers. The darkened mind however, will no more see, than the darkened heart will feel. And so the atheistic materialist looks upon the cosmos, with all its vast grandeur and majesty and declares; "Happenstance." Unto these, reality is but a brute fact, requiring neither explanation nor consideration, much less contemplation. And surely not deliberation. Unto these, the only nobility is in "measurementation".

---

[160] Barrow and Tipler, *The Anthropic Cosmological Principle* (Oxford: Oxford University Press, 1986).
[161] See Simon Conway Morris, *Life's Solutions: Inevitable Humans in a Lonely Universe*, (Cambridge: Cambridge University Press, 2005).
[162] Incredibly, the metaphysical (i.e., quantum) laws that underwrite our spiritual being, also serve to underwrite those physical laws responsible for our physical being!

But is such a worldview even plausible? Is it really reasonable to believe that this rationally intelligible universe, with its deep-seated elegance and economy, simply popped into being from absolute nothingness? Is it really credible that this universe, with its marvelously emergent laws and purposive directionality owes solely to blind process? Can we really believe that the fortuitous sea of quantum indeterminacies just happened to haphazardly generate the complex of higher laws responsible for sentient life and the human mind—particularly when that mind is uniquely endowed with a God-consciousness?

To each of these questions, the answer is a resounding no! And yet this is what atheistic materialists *must* believe! Not to be good scientists, mind you; but to be good atheists! (And hence, bad metaphysicians). So, regardless of how many times the militant anti-theists repeat it: science does not lead to atheism. Not, at least, unless you're willing to blur the line between truth and fiction and cling dogmatically to a number of wholly unwarranted, arbitrarily imposed, unscientific ideals!

Order from Disorder

So, from the collective bulk of micro-level indeterminacies there emerges the stable order of macro-level determinacies! Classical-level certainty then, ultimately emerges from the collective wash of quantum-level uncertainties.

This is only possible because the degree of uncertainty at the quantum level has been stochastically bridled in just such a way by God. For, if these stochastic constraints were suddenly lifted by God, cosmic chaos would immediately ensue, as the deep structure of the universe became literally undone. However, because these stochastical laws *do* exist, and possess the particular degree of limiting ranges that they do, the highly ordered universe we observe is possible.

Providentially speaking, God acts *specially* to determine the outcomes of particular individual quantum events, and also *generally* to sustain the lawful values of stochastic constraint—which, in turn gives rise to these lawful regularities of classical reality.[163]

---

[163] Along these lines Murphy writes: "The laws that describe the behavior of the macro-level entities are consequences of the regularities at the lowest level...the quantum level." (Murphy, "Divine Action and the Natural Order", *Chaos and Complexity: Scientific*

The *Logos*, or divine Mind appears to have guided the natural history of creation by determining indeterminacies within the abstract swarm of potentia that is the quantum underverse. At the quantum level then, God acts both *specially* (in determining particular quantum event outcomes) and *generally* (in determining the limits of randomnity in stochastical quantum law). The key then, to understanding how large-scale order could possibly arise from micro-scale disorder, lies in the realization that God set the limits of quantum chaos. Ultimately then, aspects of *both* special providence *and* general providence derive of quantum uncertainty!

Quantum indeterminacy is not simply *an* ontological opening in nature, it is *the* ontological opening in nature; the locus from whence God directs natural history, and actuates the cosmos into being.[164]

The noumenal substratum from which physical reality issues forth is the same divinely ordained medium through which God guides and directs the constituent particles of the cosmos; both "individually" *via* special providence, and in "bulk" *via* general providence. God therefore generates both the cosmos and its rationally intelligible structure from the bottom-up.

That God both creates and upholds the cosmos from the bottom-up is an extremely alluring postulate. Prior to the quantum revolution (and particularly the formulation of the Copenhagen interpretation) there was really no good way to imagine how a disembodied consciousness like God might possibly create (i.e., materialize) and sustain (i.e., continue *to* materialize) a physical universe into being, out of, as it were, nothing at all.

Perspectives on Divine Action, vol. II., ed., Robert John Russell, Nancey Murphy, Arthur Peacocke.1995).

[164] Regarding "ontic gaps" let me be clear that quantum indeterminism does not speak to the presence of a gap *in* the world's ontology, but rather of a gap provided *by* and *through* the world's ontology! That is, I am not here speaking of an 'ontic gap' in the sense of God temporarily suspending natural law; rather, I am speaking of an 'ontologically-*based* gap' in the sense of indeterminism *itself* serving as the fundamental source of ontic openness in nature (thereby providing God with the causal freedom to act without having to either suspend or violate nature's regularities). Such indeterminacy allows for the "non-interventionist objective divine action" of Robert John Russell. See, Robert John Russell, "Does 'The God Who Acts' Really Act? New Approaches to Divine Action in Light of Contemporary Science", in *Cosmology: From Alpha to Omega; The Creative Mutual Interaction of Theology and Science*, (Minneapolis; Fortress Press, 2008). From the causal insufficiency of quantum events there emerges the ontic space for God to act. God is indeed subtle. And brilliantly so!

Quantum physics not only corroborates Christian theology, *it realizes it!* For the first time in history, the foundational Christian doctrines of *creation* and *creation continua* have an empirical basis—a scientific grounding in reality.[165]

I can't imagine that this fact wouldn't give even the most ardent atheist pause.

If the universe is simply a brute fact, as the atheist would have us believe, then "brute fact" went a heck of a long way *out of its way* to make the universe "appear" as if it had been created by the Judeo-Christian God, *in the way that the Judeo-Christian God is said to have created it!*

Atheism and Theism

If Christianity is true and physicality derives of spirituality, then we shouldn't be surprised to find something like quantum ethereality existing beneath it all—that is, at the most foundational ontological level of physical reality.

On the other hand, if atheism is true, then why metaphysics at all? Why the mysterious mind-like substratum? Indeed, why "anything" that smacks of an objectively existent spiritual essence? What's more, why the incredible consonance between quantum reality and Christian revelation? These are excellent questions for which atheism has no good answers.

Conversely, the quantum-Creator-model suggested in this work is able to both handle *and explain* the data of hardcore physics—not to mention its consonance with Christian theology.

What place then, for "wave-function ethereality" in an atheistic model of reality? What place is there within an atheistic universe for an omnipresent, transcendent wraithlike medium that panders to the tenets of Christian theism? Honestly, isn't the reality of an abstract, mind-like substratum existing just beneath the world's materiality wildly

---

[165] According to the doctrine of *creation continua* (or "continuous creation") if God were to cease from His creative duties for even a moment the cosmos would instantaneously collapse away into nothingness. Creation then, isn't just some distant, remote, isolated event of the past. It's an ongoing event. A moment by moment miracle continued by God, with the world being ever actuated into existence through the transcendent medium of wave-function. The question "Why is there something rather than nothing?" should thus be followed with *"And why does this 'something' continue to exist from moment to moment?"*

out of character with the atheistic ideal of an exclusively materialistic worldview? Absolutely! Especially when it's the mind-like medium that's giving rise to the materiality itself!

For secular materialists, the discovery that matter comes from mind (and not *vice versa*) is a brutal realization. For Christian theologians however, (who have long taught that God in one way or another "cogitates" the cosmos into being) quantum theory is the strongest possible scientific confirmation that one could hope for!

So then, not only is the physical character of the universe radically incompatible with an atheist model (deriving as it does from pure metaphysics); it is gloriously compatible with a Christian theist model!— *Such as the one set forward in this work.*

Regarding the latter, (that is, "*the Christian theist model set forth in this work*"), consider how specifically it answers each of the pointed challenges laid down by University of Delaware Professor of philosophy Frank B. Dilley, who, in no uncertain terms, states:

> Defenders of miracles ought to be bringing forth real explanations, in terms of physics for example, of how it is that God acts. There should be a Christian physics…in the sense of showing in science itself how it is that God acts, where he is, what means he uses, and so on. One should be attempting to show that in some cases the actual distribution of physical forces is changed in unnatural ways.[166]

This seemingly insurmountable set of challenges has been met, each and every one, by the quantum-Christianity model set forth in this work!

Dilley here defies defenders of classical (interventionist) divine action to provide an intelligible account of their worldview—*even specifying the individual challenges that a coherent model of divine action would need to meet.*

For instance, Dilley suggests that divine action should be governed by "Christian physics." I suppose Dilley believes that Christians should

---

[166] Frank B. Dilley, "Does the 'God Who Acts' Really Act?", in *God's Activity in the World: The Contemporary Problem.* Ed. Owen C. Thomas, (Chico, CA; Scholars Press, 1983). Dilley's words here provide a nice synopsis of the various challenges set forth by contemporary atheists and liberal theologians against objective divine acts of God.

be able to identify within nature a domain where "Christian physics" is exclusively operative.

By "Christian physics" I suppose Dilley means a set of physical processes that differ from ordinary physical processes—a set of processes governed by different laws than ordinary reality—laws that uniquely pander to Christian phenomenology—particularly divine action.

But, if this is what Dilley means (as I am most certain it is), then this most spectacular challenge has been most spectacularly met by quantum physics!—*as amply demonstrated throughout this work!*

Dilley further suggests that "divine action" should be governed by a "Christian mathematics"! This is a murky claim however. For, who could possibly know what a "Christian mathematics" (i.e., a

Answer: you and I, thanks to quantum mechanics!

What is the Schrödinger equation, after all, but the mathematical means of generating descriptions of an underlying metaphysical reality! And what is a wave-function, but an objectively existent description of the abstract mathematical processes *through which God brings forth the world, and acts within it!*

Herein, that is, within these quantum processes, lie everything that Dilley demands!

Herein is the "explanation of divine action" Dilley asks for!
Herein is the "locus" and the "means" of divine action Dilley requests!
Herein are the "unnatural ways" Dilley speaks of!
Herein is the "physics-based" account Dilley demands!

Again, each of Dilley's spectacular challenges are spectacularly met by divine quantum action.[167] What's more, God does Dilley one better, by creating a set of physical laws that allow him to act in nature *without*

---

[167] Divine quantum action (DQA) looks suspiciously like the elusive Farrerian "causal-joint… between infinite and finite action" (see Austin Farrer, *Faith and Speculation: An Essay in Philosophical theology* [London: A. and C. Black, 1967]).

having to "change the distribution of physical forces" as Dilley assumes God would have to do!

All things considered, the quantum-theism model of reality presented in this work is the best model of reality available to mankind.

Chapter Seven

How to Design a Christian Universe

Clarity Seeking Light

My purposes in writing this book were not to persuade those outside the faith as to the truth of Christianity. Rather, it was written to edify those who already believe and have accepted the truth of Christianity on other grounds (e.g., faith, religious experience, the testimony of Scripture, the testimony of the Spirit, reason, natural theology, etc.). Ideally, this work is for the thinking Christian. Particularly those interested in, and/or perplexed by, the current science/faith controversy. It is for any believer hoping to reconcile their heart knowledge with their head knowledge. Such is a noble endeavor for any believer. Moreover, it's a natural endeavor for any believer. As Anselm stated a millennium ago: "*fides quaerens intellectum*"; or, "a confident faith seeks intellectual understanding of itself." This is what I have sought to do here; bring intellectual understanding to certain criteria of the faith. And thereby bring the faith to an intellectual (i.e., *scientific*) understanding of itself as objective truth within an objectively realist Christian universe! I thus hope to have in some small way, helped the reader to see Christianity as the objective reality that it genuinely is.

The Search

After many years of independent research I have come to conclude that there is much empirical data that the universe is *objectively* and *designedly* Christian—*in itself*; wholly apart from any subjective religious ideals we might personally project onto it.

This claim stands in direct contradistinction to the secular academic assertion that *all* religious experience is bogus, subjective self-deceit. However, while not '*all* religious experience is subjective self-deceit', *much still is*. The secularist assertion then, while not altogether true, is for the most part true. Simply consider; an individual born into a Muslim household in a Muslim country would most likely become a Muslim, having been indoctrinated since birth into the religion of Islam. Likewise, an individual born in Israel, into a Jewish household would most likely practice Judaism. The same goes for an individual born into either a Catholic, Protestant, or Evangelical Christian household in America.

That something as incidental as geography could figure so decisively in one's religious worldview speaks loudly to the subjective character of religious belief. The uncomfortable truth is that whether Hindu, Buddhist, or Atheist, we all come to the "nature-of-reality" question with deeply ingrained biases—*subjective biases* that have either been taught to us or are our own, based on things like personal experience, predilection, taste, etc.

So, in the face of so great a cloud of *subjective* witnesses then, who and what are we to trust? Who are we to believe? Can we even trust ourselves to make the judgment? Well, it all depends. It depends upon whether or not "*all* religion is subjective", as secularists and atheists tell us.

The answer, as you might recall, is that while there is indeed much that is personal and subjective in religion, not every religion is entirely subjective. Put differently: "much subjectivity" in religion, does not equate to "no objectivity" in religion. The question now becomes, how do we know which religion *is* objectively true? And to know this we must discover: What is the true nature of reality—*wholly apart from our personal, private, subjective metaphysical projections onto it?*

If mankind were to take off its variously colored religious spectacles what would it see?

What exactly *is* the objective status of reality? Would reality, stripped of all religious ornamentation and ideological garb, corroborate *any* of humanity's metaphysical belief systems?

Would the true, actual essence of reality as it genuinely and objectively exists 'in itself', even suggest the existence of a God? And if so, *which God*? Allah? Yahweh? Or perhaps one or all of the 330 million Hindu deities? Or would it be the Trinitarian God of Judeo-Christianity, who, through Christ, personally involved Himself in the eternal redemptive opera of creation, as revealed in Scripture?

Until recently, it seemed the answer to this ultimate question of questions would forever elude us. As it turns out however, (and as we have seen) the structure of this physical world allows us to probe it in very revealing ways. Utilizing ultra-modern technology, physicists can now, and for the first time in scientific history, empirically explore reality at literally its most elemental level; that being, the level of subatomic reality and beneath.

Quantum physics has allowed mankind to pierce the veil of deep reality; to glimpse what lies beneath the physical garment that is this material world. And beneath it all, that is, beneath the most elemental constituents of this material world, science has come face to face with naked reality! And it is unlike anything secular science would have ever imagined!

As it turns out, naked reality is a profoundly metaphysical reality; governed by profoundly metaphysical phenomenon. It is a mind-like reality; and yet it is the mother-substrate of physical creation. Naked reality is an objectively ethereal realm. This ethereality however, is more than a mere substrate; *it is an ontic nexus*! It is in fact *the* ontic nexus from whence the whole of physical creation derives. Hence, if quantum physics has taught us anything, it is that there is indeed a Final Frontier—and it is *not physical!*

Here is a spiritual reality that transcends physical reality, just as the majority of the world's religions have long taught. Quantum physics thus corroborates the general religious assertion of a transcendent metaphysical realm. Quantum physics can thus be said to support a large variety of very general religious views. And indeed quantum physics is said to support a large variety of very general religious views.

This is one reason why quantum physics has been so popular with postmodernists. You see, the notion of 'generalized spirituality' fits perfectly with the postmodern ideal of variegated spirituality (the ultimate goal of which, being, to rid the world of all absolutes and provide in its place a bland but pious gruel, served lukewarm for the deadened masses to subsist on). Fortunately however, 'transcendence' is not the last word from quantum physics. For, upon closer inspection, the self-same quantum ethereality associated with quantum transcendence can be seen to underwrite some very particular religious phenomenon—namely, that associated with Judeo-Christian doctrine.

For, as it turns out, quantum reality is ideally suited to serve as an ontological foundation for Christianity, having been phenomenologically customized as it were, to realize key tenets of the Faith. This is a very telling state of affairs, in that an authentic Christian reality would demand, indeed *necessitate*, a highly unlikely world ontology. One not unlike that present at the quantum level!

Consider just how clever quantum reality truly is. Not only does it "naturalistically" generate the regular, ordinary, mundane sphere of macro-reality, but it does so from a place that is equally and oppositely *irregular, extraordinary*, and *supramundane*, albeit, in a way perfectly suited to clandestinely realize all the doctrinal phenomenon of a realist Christian reality! Now *that's quite a trick!* Even for a divine engineer *that's quite a trick!*

Remember, God must keep His ways hidden in order to insulate mankind from the epistemic pitfalls associated with seeing such a God too clearly. For, if God were too openly known, and/or too unambiguously manifest, mankind's acceptance of Him would be reduced to a mere intellectual exercise, in turn spoiling the faith-based return God seeks through Christ. In essence, the private, personal, heart-felt truth of Christ would be replaced by a dry, civic, head-based recognition of Deity.

Physics and Metaphysics

The metaphysical implications of quantum physics should be undeniable to any unbiased person. The facts are simply the facts: we exist atop a ghost-world! This, again, is bad news for the materialist. And it gets even worse, in that we not only exist "atop" a ghost-world, but also, *by and through* a ghost-world! Whether atheistic materialists like it or not, metaphysics is quite *literally* part of the world ontology!

Quantum reality unobserved is a fundamental, indeed, *elemental* reality that is in no way physical. Ontologically speaking, it is metaphysical, plain and simple. What's more, it is metaphysical in a uniquely "mentalistic" way; such that the only thing even remotely like it in all human experience is consciousness. We thus appear to be dealing with an ethereality not unlike a "disembodied consciousness."[168]

Quantum reality however is more than just "mind." Indeed, it is more than just "spirit." It is, in its fullest sense: a *soulish reality!* And, like most souls, it lay hid beneath the material cloak of physical reality. As stated above, this *naked* reality, is *unlike* anything previously imagined in science. At the same time however, this same *naked* reality is extraordinarily *like* everything previously imagined in theology!

The End of Scientific Materialism

The present-day scientific paradigm is a materialistic paradigm. And it governs according to the dictates of four cardinal precepts: naturalism, atomism, reductionism, and determinism. These four precepts have ruled mainstream science for almost three centuries in their fullest ontological forms. Today however, this materialistic reign of hopelessness is over! For, the new *quantum ontology* has presented us with a far clearer picture of ultimate reality, while effectively falsifying each of the above ontic precepts!

Which, precepts, when taken together, leave little room for deity, *and no room for the Christian God.* Within such a paradigm atheism is taken to be a foregone conclusion; God a delusion, and free will an illusion. Such is the nature of the modern materialist scientific paradigm. Which, paradigm, reigned supreme for almost three centuries before being brought to its crushing end by the quantum revolution. (A revolution begun by Max Planck in 1900 and completed by Niels Bohr and Werner Heisenberg in 1927).

Since the quantum revolution, the tables have been turned. And this time "ontology" is in the theist's corner! Quantum mechanics has greatly expanded our conception of reality. And having done so, has greatly upset the balance of the science/theology debate. Materialism, freshly assessed against the penetrating insights of quantum physics, has proven to be

---

[168] Einstein himself disparaged quantum mechanics (particularly non-locality) refusing to believe that God would use "telepathic methods." (See Einstein's letter to Cornel Lanczos, March 21, 1942). Of course, today we know that Einstein was exactly wrong about quantum non-locality; for in 1983 Alain Aspect experimentally proved the phenomenon. The question now is what to make of this apparent "telepathy" between particles?

little more than a groundless philosophical refuge for atheist's, having nothing at all to do with either legitimate science or genuine reality.

Christian theism on the other hand, when assessed against these same 'penetrating insights', appears radically and *ontologically* suited for realization! Quantum physics then, has not only *refuted* scientific materialism, but also *corroborated* Christian theism! Consider: the novel ontology of quantum mechanics has single-handedly falsified each of the underwriting tenets of scientific materialism![169] (Quite a feat when you consider that scientific materialism has been the prevailing scientific paradigm for the past three centuries!). More incredible still, is the fact that the 'metaphysicality' quantum physics brings to the scientific table *just happens to cater quite thoroughly to the ontologically-based tenets of Judeo-Christianity!*

This is a profoundly deep truth about the ultimate nature of reality. One that eclipses every other scientific discovery ever made! Science must now deal with the metaphysical. For, according to science, we live in a metaphysical world. The next step for science then, is to admit the blunder of materialism, acknowledge the reality of matter/Ψ dualism, and move forward with a newfound appreciation for metaphysics. A "metascientific" paradigm, it seems, is inevitable.

Material monism is dead: that which is material, originates out of that which is immaterial. Still, such a paradigm is certain to be vehemently resisted by the materialist paradigm in place. Any materialistic resistance however, will be futile. For, it has now been empirically established that the ontological underpinnings of reality are metaphysical; perhaps even spiritual—not to mention religiously implicative! Quantum reality after all, is phenomenologically suited to realize the very particular theological tenets of a very particular Deity. Namely, *the "God" of Judeo-Christianity!*[170]

---

[169] Which tenets, were cultivated out from the erroneous ontology of classical mechanics.

[170] The quantum-based Christian tenets spoken of throughout this work, rather than being historically-based (such as the death, burial, and resurrection of Jesus Christ), are ontologically and/or existentially-based, rendering them therefore open to scientific verification and/or falsification. For instance: *Free will* necessitates a "causally indeterministic" ontology; *Special providence,* an "open" ontology (which also preserves the integrity of general providence); *General providence,* an "immutable regularity of natural laws and processes" (which in turn preserves the ontological openness requisite to special providence); *Creation ex-nihilo,* a "finite, contingent universe"; *Creation continua,* a "non-static, dynamic, contingent universe that is in a constant state of 'becoming' and/or 'coming into being' as it were, from nothing"; etc. We thus discover God in the shrewdest

Of course, the blatantly theistic implications of the new dualist quantum ontology doesn't remotely jibe with the atheistic tenor of the contemporary materialist paradigm of science. Much to the dismay of the latter, scientific advance *has not* demonstrated once and for all that "God is dead!" In fact, if science has demonstrated anything, it is that God is very much *alive*; very much *interested*; and very much *involved* in the human drama. This much can be inferred from the ontic character and structure of the universe; which, as we have seen, is set-up to realize a very specific theological detail; namely, that set forth in the Judeo-Christian Scriptures!

It is important to understand exactly what God has done here. God has engineered a universe that is intrinsically and characteristically "Christian." And has done so in a most elegant and economical manner, incorporating but a single, subtle, near infinitesimal physical constant; h, otherwise known as the Planck constant.

Understand, creating and/or engineering a universe such as this would have been a very tall ontological order. So extreme would have been the creative/engineering demands of such a reality that nothing less than an omnipotent, omniscient Creator/Engineer would have possibly sufficed.

The significance of this cannot be overstated. Christian doctrine is not some vague set of meta-principles so general in character that they can be massaged to accommodate any worldview. The tenets of Christianity are very definitive, requiring an equally definitive ontological footing. Quantum ontology realizes this footing, being spectacularly geared towards theological economy, elegance, and utility.

This fact (and it is a fact) is highly suggestive. And what it suggests is a very simple truth: namely, that *Jesus Christ is Creator, Lord, and Savior of the world*. Indeed, unlike the innumerable man-made religious cults that have risen and fallen across the ages, the Judeo-Christian faith is here to stay. For, it is woven into the very fabric of reality.

Final Thoughts

The reality we find ourselves occupying seems to be informing us that Christianity *is definitely not* simply one more man-made religion

---

of all locations: beneath and behind the world's ontological foundations! Shrewder yet, even when He is found, He is not found! Faith thus abides!

amongst the world's myriad of socially constructed, culturally subjective, belief systems, but is rather, an objectively existent truth grounded in the very ontology of reality itself.

With a single quantum principle God has managed to realize every major ontic-based doctrine, tenet, and truth of the Judeo-Christian faith. Such as • Free will (and the requisite volitional realm to realize it) • Special providence • General providence • *Creation ex-nihilo* • *Creation continua* • The doctrine of the Soul •   Matter/Spirit dualism • Divine hiddenness • *and various others.*

More stunning still, is the fact that the unique nature of the quantum ontic underwriting each of these theological requisites allows for their robust realization, yet without ever once coming into conflict with the scientific tenets of methodological naturalism! Quantum physics thus reveals how the metaphysical tenets of Christianity can exist side by side with naturalistic science, *without ever once overstepping a naturalistic boundary!*

So beneficial is the quantum ontic to realizing the metaphysics of Christian theism, that a quantum ontology may literally be the difference between a *theistic* universe and an *atheistic* universe! (The latter, e.g., being of the Laplacian sort) If there is any such thing as a *Christianity constant* then, it is the Planck constant! (Which, recall has the near infinitesimal value of $6.62606896(33) \times 10^{-34}$ J·s.). Never before has an engineer been able to accomplish so much with so little. Only a designer the caliber that of the Judeo-Christian God could have wrought such a profoundly economical solution to the longstanding science/faith conflict. At the end of the day then, what science takes to be the mysterious nature of quantum mechanics is but the subtlety of God realizing His divine plan.

You believe that the world is not subject to the accidents of chance, but to divine reason. Therefore, you have nothing to fear. From this tiny spark, the living fire can be rekindled.

—Lady Philosophy to Boethius

Index

A

Absolute Creationism, 118
abstract objects, 4, 84, 93, 117
    objective existence of, 93, 113
abstract wave-function realism, 35,
    40, 54, 62
a-causality, 13, 15-17
Allah, 132
anthropic fine-tuning, 122, 148
anti-realism, 43, 45-49, 52-53
Aquinas, Thomas, 91, 150
Aristotle, 3, 8
Aspect, Alain, 43, 96, 134
Athanasius, 45
atheism, xiv, 5, 124, 126
Augustine, xiv, 80, 83, 118

B

Barbour, Ian, 34, 120
Barrow, Isaac, 44
Bauer, Edmond, 108
Beck, Friedrich, 14
Behe, Michael, 99

Bell, John, 45
Big Bang, 83, 118
Boethius, 118
Bohm, David, 45
Bohr, Niels, 5, 10, 40-43, 45-46,
    48, 50, 53, 62, 71, 86, 96,
    134
Boltzmannian statistics, 121
book of nature, 55, 80
book of Scripture, 55, 80
Born, Max, 49, 86, 106
Bose-Einstein (BE) statistics, 121
Boyle, Robert, 44
brain-body interface, 38
Broglie, Louise de, 13, 113

C

Cabanis, Pierre, 78
Cartesianism, xiv, 38
causation, xx, 36, 66, 69-70, 89, 92,
    94
Chalmers, David, 15, 62, 78, 110
Christ, 23, 28, 44-45, 53-54, 56,
    58-59, 61, 73, 92, 132-33,
    135-36

Christianity, xiv-xv, xxi-3, 7, 10, 14, 16, 21-22, 24, 30, 35, 38, 52, 56, 69, 74-75, 79, 90, 104, 114, 116, 118, 126-27, 130, 132-33, 135-37
   Catholic, 91, 131
   Protestant, 24, 131
Christianity constant, 137
clockwork universe, xx
Colbeck and Renner, 113
Colyvan, Mark, 93
Como, Italy, 41
complementarity, 41-42, 45
Compton, Arthur, 52, 81, 106, 120
conceptualism, 118
Connes, Alain, 6, 86
consciousness, xviii, 5, 9, 14, 52, 62, 66-67, 70, 72, 78, 86, 89, 107-11, 122, 124-25, 134
Copenhagen interpretation, 42-46, 52, 63, 72, 78, 91, 125
Copenhagenism, 43-44, 48
Coyne, George V., 34
creation continua, 10, 13, 83, 88, 118, 126, 135, 137
creation ex-nihilo, 135, 137

D

Davisson and Germer, 13
deism, xiv, 13, 25, 30-31, 96
delayed-choice experiment, 6, 11
Descartes, René, xx, 13, 118
determinism, xiv, 5, 19, 30-31, 33, 61, 75, 89, 121, 134
   ontological, 30-31
Dilley, Frank B., 127-29

divine action, xiii-xv, xvii, xxi, 16, 21, 24-27, 29-36, 39, 45, 81, 89, 91, 96-97, 100-103, 105-6, 119, 124-25, 127-28
   interventionist, xiv, 35, 98, 119, 125, 127
   mechanics of, 36, 39, 100
   non-interventionist, 35, 98, 119, 125
   within a law-bound world, 21
divine Mind, xviii, 9-10, 20, 26, 37, 51, 56, 58, 60, 66-68, 72, 84, 90, 103, 110, 118, 121, 125
divine purpose, 10, 36, 83
divine will, 17, 20, 31, 90-91, 103
DNA (deoxyribonucleic acid), 37-38
double-slit experiment, 6, 51, 82, 90
DQA (divine quantum action), 89, 91, 97, 128
dualism, 9-10, 13-15, 27, 35, 74-75, 77, 111, 119, 135
   mind/matter, 13
   ontological, 9, 13, 35, 74
   quantum, xvii, 8, 10, 14, 27, 36, 42, 64, 72-74, 80, 82, 86, 112
   wave-particle, 14, 54
Dyson, Freeman, 9

E

Eccles, John, 14-15, 150
Eddington, Arthur, 5, 71
Einstein, Albert, 3, 5, 7, 13, 43, 57, 83, 86, 94, 96, 106, 113, 134, 151-52, 157
Enlightenment science, xx, 65
epiphenomenalism, 78
Erasmus, 44

evolution, 11, 14-15, 57-58, 67, 77, 85, 98-99, 113, 123

F

Farrer, Austin, 91, 128, 151
Fermi-Dirac (FD) statistics, 121
Feynman, Richard, 73, 82
Fictionalism, 117
Foster, John, 15
free will, xiv, xx-xxi, 10, 14, 21, 32, 53, 81, 106, 112, 119, 134-35, 137
Frege, Gottlob, 6
Freud, Sigmund, 22

G

gaps, xiv, 16, 25, 31, 89, 98, 100, 125
   ontic, xviii, 6, 8, 10, 15-16, 26, 38-39, 42, 48, 51, 53, 72, 75, 80, 106, 114, 116, 125, 132, 134, 136-37
Gilkey, Langdon, 45
God, xiii-xviii, xx-xxi, 2, 6, 10, 13-15, 17-22, 24-39, 45, 52, 54-58, 60-62, 64, 66-68, 72-77, 79-80, 83-84, 89-92, 95-106, 108-10, 115-29, 132-37
   as determiner of indeterminacies, 19, 89, 97, 105, 110
Gödel, Kurt, 6
god-of-the-gaps, xiv, 31, 98
God's two books
   quantum unification of, 80
Grangier, P., 43

H

Hardy, Alister, 14
Hardy, G. H., 86
Heim, Karl, 52, 81, 106, 120
Heisenberg, Werner, 42
hidden variables, 43, 45, 91
Hinduism, 131-32
Hodgson, David, 15
hope, xvi-xviii, xx, xxii, 22-24, 28, 34, 50, 103, 113, 127, 130
   Christian, xxi-1, 3, 5-6, 8, 10, 15, 20-29, 32, 34-36, 39, 45, 48, 53-56, 58-59, 61, 64, 66, 68, 74-75, 80-81, 89, 94, 97, 102, 104-6, 109, 112, 114-20, 123, 126-28, 130-31, 133-37

I

indeterminacy, xxi, 14-17, 35, 42-43, 54, 81, 89, 91, 120-21, 125
   as cloak for the divine hand, 72, 118
Indispensability Thesis, 93
instrumentalism, 45
intelligent design, 37, 98-99
interfacing, 14, 21, 29, 32, 34, 38, 104, 110
   divine mind-to-wave-function, 19
   God-to-world, 97
   mind-to-brain, 14
   soul-to-brain, 14
   soul-to-wave-function, 90
   spirit-to-matter, 29, 34
interference, 6-7, 40, 48, 50-52, 63, 70, 93, 114
   constructive, 23, 63, 70

deconstructive, 63, 70
Islam, 131

J

Jackson, Frank, 14
Jammer, Max, 63
Jesus of Nazareth, xvi, 22, 25
Job, 22
Judaism, 6, 131
Judeo-Christianity, xiv-xv, 3, 10, 116, 135

L

Laplace, Pierre-Simon, xiv, 25, 27, 30-31, 79
Leibniz, Gottfried Wilhelm, xiv, 31, 118
Leslie, John, 6
locus of providence, 28
Logos, 20, 53, 56, 58, 61, 77, 79, 84, 125
London, Fritz, 108

M

Margenau, Henry, 14
materialism, xxii, 5, 19, 50, 69, 92, 116, 134-35
   scientific, 5, 134-35
mathematics, 35, 40, 55, 57-58, 66-67, 107, 148-49, 152, 155-57
   higher, 57
measurement, 42, 46, 53, 64, 88, 108, 111, 114
Mechanism of Divine Action, 96
Mermin, N. David, 71-72, 108

metaphysicality, 70, 73
metaphysicians, 82, 92, 110
metaphysics, 21, 25, 54, 81, 126, 133, 135, 137, 147, 152, 158-59
middle-knowledge, 66, 68, 74, 81, 85-86, 88, 91-92, 118
Mind of God, 13, 17, 20, 36-38, 56, 77, 90, 102
Molina, Luis de, 85
monism, xvii, 5, 46, 78, 135
   material, xiii, xv, xvii, xx-xxi, 4-5, 7, 9, 12, 15, 20-21, 27, 29, 32-33, 35-37, 40, 42, 46-47, 50, 53, 62, 64, 66, 68, 70, 72-78, 83, 86, 92, 95, 98, 100-102, 105, 107-8, 112, 117, 121, 132, 134-35
Murphy, Nancey, 34-35, 89, 125

N

naturalism, 24, 50, 134, 154
natural law, xiv, xviii, 24, 35, 98-99, 119, 125, 135
Neo-Meinongianism, 117
Neo-Platonism, 6, 44, 111
Newton, Isaac, xiv, 19, 30-31, 34, 44-45, 57, 70
Newtonianism, xiv
Niebuhr, Reinhold, 21
nominalism, 93, 117
nomology, 3
non-locality, 15, 35, 43, 45, 68, 134
noumenality, 20, 72

O

omnipotence, xvi
omnipresence, 68, 74, 94
omniscience, 68, 74, 85, 94
ontology, 3, 19, 31-32, 45, 47, 49, 56, 74-76, 106, 120, 134-35, 159
    abstract, 61
    Bohr/Heisenberg, 10, 43, 50
    Christian, 26-27, 36, 58, 64, 74, 80, 102, 106, 114, 120, 136
    dualist, 5
    indeterministic, xxi, 135
    platonic, 6-7, 48
    quantum, xv, 46, 65, 69, 72, 74-76, 80, 117, 134, 136-37
Origen, 44

P

Paley, William, 100
Paul (apostle), 28
Penfield, Wilder Graves, 14
Penrose, Roger, 6, 52, 54, 122, 153-54
philosophers, 2, 4, 14, 78, 82, 84, 86, 92, 110, 112, 114, 149
philosophes, 25, 44
physicalism, 44, 48, 53, 112
physicality, xv, xx, 5, 17, 26, 42, 62, 68-70, 73, 76-79, 86, 92, 98, 107, 115-16, 126
physics, Newtonian, xiii, xix, xxi, 25, 42, 44
Planck, Max, 86, 115, 134
Planck's constant, 42
Plato, 3-6, 44, 54, 79, 86, 92
platonism, 4, 6, 10, 44, 86, 111
Plotinus, 6, 44, 86
Polkinghorne, John, 35, 47, 68, 80, 82, 155, 157
Pollard, William, 45, 81-82, 106, 120, 155
Popper, Karl, 14, 54
positivism, 45, 47
postmodernism, xxi, 27, 29, 37, 69, 71, 82, 92-93, 114-15, 122, 132, 151-52, 157
potentiality catalogue, 8
probabilistic interpretation, 50, 62
providence, xiv, xxi, 10, 28-29, 31-33, 36, 39, 60-61, 81, 89, 98, 100-101, 121, 125, 135
    general, xiv, 10, 24, 33-34, 43-44, 56, 61-62, 73, 92, 125, 132, 135-36
    special, xiv, 10, 24, 29, 33, 60-61, 125, 135
psi, 47, 49, 61

Q

quanta, 2, 8, 18, 111
quantum cosmology, 34, 83, 156
quantum divine action, 35
quantum entanglement, 68, 94
quantum essence, 4, 90, 95, 111
quantum indeterminacy, xxi, 14-17, 19, 35, 42, 54, 89, 91, 97, 120, 124-25
quantum indeterminism, 15, 27, 35, 37, 45, 50, 63, 125
quantum mechanics, xvii-xviii, xxi-2, 5, 7, 9-10, 14, 17, 21, 25, 27, 30, 32, 35, 39, 41, 45, 47, 54, 56, 62-63, 65, 74-75,

81-83, 92, 94, 97, 100, 106-7, 110, 113-15, 122, 128, 134-35, 137
  laws of, xviii, 9
  suited to Christianity, 26
quantum metaphysics, 80, 93, 110, 114
quantum non-locality, 15, 45, 134
quantum ontology, i, iii, xv, xxii, 2, 7, 9, 12, 15, 21, 35, 39, 43, 46, 49, 55-56, 65, 69, 72, 74-77, 80, 82, 88, 93, 105, 108-9, 117, 123, 134, 136-37, 149, 152, 156
  suited to resolve theological paradoxes, 120
quantum particles, 2, 8-9, 14, 16, 60, 62, 66, 84
  physical, 8, 40, 60
quantum phenomenology, 26, 35, 40, 95, 97
quantum physics, ix, xvii, xxi, 1-3, 5, 9, 13, 19, 21, 24, 27, 35, 38-39, 41-42, 47, 54, 64, 73-74, 77-83, 86, 88, 91, 93-95, 97, 106-8, 111-12, 114-16, 118-20, 126, 128, 132-35, 137, 150, 156
  contemporary, xxi, 118
quantum process, 4, 15, 24, 26-27, 39, 61, 72, 94, 109, 112, 114, 118, 121, 128
quantum reality, xxi, 3-6, 11, 13, 16-17, 20, 24, 27, 34, 36, 38-40, 42, 44-48, 50-53, 55-58, 60-61, 63, 65-66, 68, 70-73, 75-76, 83, 85, 89, 93, 95, 97, 105-11, 113, 115, 121, 126, 133-35
  discovery of, 66, 70
  ontic structure of, 38-39
quantum revolution, xv, xvii, 21, 30, 65, 69, 92, 95, 98, 100, 115, 125, 134
quantum systems, 9, 14, 16, 18, 41-42, 46, 48
quantum theory, 3, 5, 9, 11, 15, 17, 25, 35, 40, 43-44, 47, 51, 62, 69, 78-83, 91, 97, 107-8, 113-15, 120, 127, 148, 150-51, 153
quantum void, 12, 76-77
quantum wave-functions, 4-12, 14-15, 17-19, 35, 40, 49, 60-63, 66, 69, 72, 74, 83-85, 106, 109, 113-14, 118
quantum waves, 9, 14, 62
Quine, W. V. O., 6

R

realism, 45, 48, 52-53, 62, 117
  Christian, 21, 59
  critical, 44, 46-48
  mathematical, 10
  platonic, 40-41, 46, 48-50, 63, 91, 117
  quantum, 48, 50-53, 55-58, 61, 63, 65-66, 68, 70-73, 76
  spiritual, 1, 3, 6, 8, 14, 16, 20-22, 26-27, 32, 41, 44-46, 48-49, 51-52, 55, 64, 69-70, 73, 75-79, 95, 97-98, 102, 104, 106, 114, 116, 122-23, 126, 132, 135

wave-function, 35, 40, 49-50, 54, 62, 91, 113, 117
reductionism, 5, 19, 33, 78, 112, 134
regularities of nature and science, xix, 55, 89, 119, 125
Robinson, Daniel N., 14
Roger, G., 43
Russell
   Bertrand, 6, 66
   Robert John, 34-35, 80, 89, 91, 120-21, 125

S

Sagan, Carl, xix
salvation, 26
SchrÖdinger, Erwin, 7, 47, 86, 152
science/Christianity controversy, 130
science/God controversy, 22, 29
Scripture, xv, xvii, 24, 28-29, 44, 55, 73-75, 80, 103, 109, 116, 120, 130, 132
   Judeo-Christian, 24, 74, 105, 136
Searle, John, 78
soul, xx-xxi, 6, 10, 14, 19-21, 25, 52, 54, 78, 90, 95, 109, 111-12, 115-16, 122
sovereignty, xiv, 21, 89, 117-18, 120
Sperry, Roger Wolcott, 14
spirit, xviii, xx-xxi, 12, 21, 29, 32-35, 37, 41, 45, 52, 54, 73, 78, 83, 89-90, 92, 94, 98, 100, 102, 104, 106, 109, 114-16, 120, 130, 134
spiritron, 104-5
spooky action at a distance, 94
Stapp, Henry, 52
stochastic process, 61
Stoeger, William R., 34-35
stoicism, 20

T

Tegmark, Max, 6
theism, xiv-xv, 3, 27, 105, 118-19, 126
   Christian, xxii, 3, 15, 24, 27, 32, 45, 94, 106, 109, 112, 114-15, 117, 120, 123, 126, 135, 137
theodicy, 3, 61, 89
theologians, 2, 23, 81-83, 85, 92, 99-100, 110
theology, 3, 10, 21, 24, 26-27, 31, 34-35, 38, 44-45, 61, 64, 68, 80-82, 87-89, 91, 104-5, 109-10, 114, 118-19, 121-22, 125-26, 128, 130, 134
   Christian, 26-27, 61, 64, 80-81, 89, 104, 118, 126
   conservative, 98, 110
   liberal, 24, 44-45, 74, 127
   neo-orthodoxy, 45
Thomism, 3
Thom, René, 6
Timaeus, 5-6
Tracy, Thomas F., 35, 89, 119
transcendence, 10, 18, 38, 70, 74, 76-77, 133
Trinity, 44-45, 68
tripartite ontology, 54
two-books metaphor, 55

U

uncertainty principle, 11, 42-43
unification of
    physics and theology, 3, 68, 82
    science and Christianity, 10
Universal Possibilism, 118

V

virtuosi, 44
volition, xxi, 10, 12-13, 17, 19, 53, 90
Voltaire, 25
Von Neumann, John, 62, 86, 107-8, 110, 158

W

wave-equation, 2, 4, 7, 11, 15, 18, 46, 50, 61
    solution to, 2, 88, 137
wave-function, 2, 5-9, 11-20, 35, 40, 43, 46-51, 53, 60-62, 66-67, 69-70, 72-73, 77, 81-91, 93-94, 97, 107-8, 110-13, 117, 126, 128
    as bridge between worlds, 20, 80
    collapse of, 11, 17, 67, 88, 91
    evolution of, 14, 113, 123
    as locus of divine interfaction, 32, 79, 91
    as locus of soul-brain interfaction, xx
wave-function realism
    non-physicalistic, 46, 48
    physicalistic, 14, 18, 43, 46, 48, 105, 114, 116
Wheeler, John, 6, 71
Wigner, Eugene, 55, 58, 62, 86, 89, 107-8, 110, 120, 158
Wolf, Christian, 14
Word of God, 73
world-soul, 20

Y

Yahweh, 132

Z

Zeno, 44

Bibliography

Albert, David Z, *Quantum Mechanics and Experience* (Massachusetts: Harvard University Press, 1992).

Aquinas, Thomas, *Summa Contra Gentiles: Book Two: Creation*, translated by James F. Anderson (Notre Dame: University of Notre Dame Press, 1992).

Aristotle, *Metaphysics*, translated by Hugh Lawson-Tancred (London: Penguin Group, 1999).

Armstrong, D.M., *Universals: An Opinionated Introduction* (Boulder: Westview Press, 1989).

Armstrong, D.M., *Universals and Scientific Realism*, vol. 1., (Cambridge: Cambridge University Press, 1980).

Armstrong, D.M., *Universals and Scientific Realism*, vol. 2., (Cambridge: Cambridge University Press, 1980).

Augustine, *City of God*, translated by Henry Bettenson (London: Penguin Group, 2003).

Augustine, *Confessions*, translated by Henry Chadwick, (Oxford: Oxford University Press, 1992).

Baggett, David, and Wallis, Jerry L., *Good God: The Theistic Foundations of Morality* (New York: Oxford University Press, 2011).

Baggett, Jim, *A Beginners Guide to Reality: Exploring Our Everyday Adventures in Wonderland* (New York: Pegasus Books, 2005).

Baggett, Jim, *Beyond Measure, Modern Physics, Philosophy and the Meaning of Quantum Theory* (New York: Oxford University Press, 2004).

Balaguer, Mark, *Platonism and Anti-Platonism in Mathematics* (Oxford: Oxford University Press, 1998).

Barbour, Ian, *Issues in Science and Religion* (San Francisco: Harper, 1966).

Barbour, Ian, *When Science Meets Religion: Enemies, Strangers or Partners?* (San Francisco: Harper, 200).

Barbour, Ian, *Models, Myths, and Paradigms: A Comparative Study in Science and Religion* (New York: Harper and Row, 1976).

Barbour, Ian, *Religion and Science: Historical and Contemporary Issues* (San Francisco: HarperCollins, 1990).

Barrett, Jeffrey A., *The Quantum Mechanics of Minds and Worlds* (New York: Oxford University Press, 1999).

Barrow, John, *Pi in the Sky: Counting, Thinking and Being* (Oxford: Oxford University Press, 1992).

Barrow, John, and Tipler, Frank, *The Anthropic Cosmological Principle* (Oxford: Oxford University Press, 1986).

Bell, J. S., *Speakable and Unspeakable in Quantum Mechanics* (Cambridge: Cambridge University Press, 1987).

Bohm, David, *Wholeness and the Implicate Order* (London: Routledge, 1980).

Bohr, Niels, *Atomic Physics and Human Knowledge* (London: Wiley, 1958).

Bohr, Niels, *The Philosophical Writings of Niels Bohr*, vol. 1., *Atomic Theory and the Description of Nature* (Woodbridge: Oxbow Press, 1987).

Bohr, Niels, *The Philosophical Writings of Niels Bohr*, vol. 2., *Essays 1933 to 1957 on Atomic Physics and Human Knowledge* (Woodbridge: Oxbow Press, 1987).

Bohr, Niels, *The Philosophical Writings of Niels Bohr*, vol. 3., *Essays 1958-1962 on Atomic Physics and Human Knowledge* (Woodbridge: Oxbow Press 1995).

Born, Max, *The Bohr-Einstein Letters* (New York: Walker, 1971).

Camilleri, Kristian, *Heisenberg and the Interpretation of Quantum Mechanics: The Physicist and Philosopher* (New York: Cambridge University Press, 2009).

Clayton, Philip, *Mind and Emergence: From Quantum to Consciousness* (New York: Oxford University Press, 2004).

Colyvan, Mark, *The Indispensability of Mathematics* (New York: Oxford University Press, 2001).

Copan, Paul and Craig, William Lane, *Creation out of Nothing: A Biblical, Philosophical, and Scientific Exploration* (Grand Rapids: Baker Academic, 2004).

Davies, Paul and Gribbin, John, *The Matter Myth: Dramatic Discoveries that Challenge our Understanding of Physical Reality* (New York: Touchstone, 1992).

Davies, Paul, ed., *The New Physics* (Cambridge: Cambridge University Press, 1989).

Dawkins, Richard, *The God Delusion*, (Great Britain: Bantam Press, 2006).

Dennett, Daniel C., *Breaking the Spell: Religion as a Natural Phenomenon*, (New York: Penguin Group, 2007).

D'Espagnat, Bernard, *In Search of Reality* (New York: Springer-Verlag, 1983). D'Espagnat, Bernard, *On Physics and Philosophy* (Princeton: Princeton University Press, 2006).

D'Espagnat, Bernard, *Reality and the Physicist: Knowledge, Duration and the Quantum World* (New York: Cambridge University Press, 1989).

D'Espagnat, Bernard, *Veiled Reality: An Analysis of Present Day Quantum Mechanical Concepts* (New York: Addison Wesley, 1995).

Deutsch, David, *The Fabric of Reality* (New York: Penguin, 1998).

Dirac, P. A. M., *Lectures on Quantum Mechanics* (New York: Yeshiva University, 1964).

Dodds, Michael J., *Unlocking Divine Action: Contemporary Science and Thomas Aquinas* (Washington D.C.: Catholic University Press of America, 2012).

Eccles, John C., *Evolution of the Brain: Creation of the Self* (New York: Routledge, 1989).

Eccles, John C., *How the Self Controls Its Brain* (Berlin: Springer-Verlag, 1994).

Edwards, Denis, *How God Acts: Creation, Redemption, and Special Divine Action* (MN: Fortress Press, 2010).

Ellis, George, *Before the Beginning: Cosmology Explained* (New York: Bowerdean Publishing, 1994).

Farrer, Austin, *Faith and Speculation: An Essay in Philosophical theology* (London: A. and C. Black, 1967).

Feynman, Richard P., QED: The Strange Theory of Light and Matter (Princeton: Princeton University Press, 1985).

Fine, Arthur, The Shaky Game: Einstein Realism and Quantum Theory (Chicago: University of Chicago Press, 1996).

Folse, Henry J., *The Philosophy of Niels Bohr: The Framework of Complementarity* (Netherlands: North-Holland Physics Publishing, 1985).

Foster, John, *The Immaterial Self: A Defence of the Cartesian Dualist Conception of the Mind* (London: Routledge, 1991).

Gilder, Louisa, *The Age of Entanglement: When Quantum Physics was Reborn* (New York: Random House, 2009).

Gould, Paul M. ed., *Beyond the Control of God? Six Views on the Problem of God and Abstract Objects* (New York: Bloomsbury Academic Publishing Inc., 2014).

Hebblethwaite, Brian, and Henderson, Edward, *Divine Action: Studies Inspired by the Philosophical Theology of Austin Farrer* (Edinburgh: T and T Clarke, 1990).

Heim, Karl, *Christian Faith and Natural Science* (New York: Harper and Brothers, 1953).

Heisenberg, Werner, *Physics and Philosophy: The Revolution in Modern Science* (New York: Harper Perennial, 2007).

Heisenberg, Werner, *Uncertainty: The Life and Science of Werner Heisenberg* (New York: Bellevue Literary Press, 2009).

Heisenberg, Werner, *The Physical Principles of the Quantum Theory* (Chicago: University of Chicago Press, 1930).

Hodgson, David, *The Mind Matters: Consciousness and Choice in a Quantum World* (New York: Oxford University Press, 1993).

Hofstadter, Dougas, *Gödel, Escher, Bach: The Eternal Golden Braid* (New York: Harper and Row, 1979).

Hume, David, *A Treatise of Human Nature* (Oxford: Oxford University Press, 1967).

Jackson, Frank, *Perception* (Cambridge: Cambridge University Press, 1977).

Jammer, Max, *Einstein and Religion* (Princeton: Princeton University Press, 1999).

Jammer, Max, *The Philosophy of Quantum Mechanics* (New York: Wiley, 1974).

Jauch, J. M., *Are Quantum Real? A Galilean Dialogue* (Indianapolis: Indiana University Press, 1973).

Jeans, James, *Physics and Philosophy* (New York: Dover, 1981).

Kant, Immanuel, *Critique of Pure Reason*, translated by Marcus Weigelt (New York: Penguin, 2007)

Kosso, Peter, *Appearance and Reality: An Introduction to the Philosophy of Physics* New York: Oxford University Press, 1998).

Kragh, Helge, Quantum Generations: *A History of Physics in the Twentieth Century* (Princeton: Princeton University Press, 1999).

Krauss, Lawrence, *A Universe from Nothing: Why There Is Something Rather than Nothing*, (New York: Free Press, 2012).

Kumar, Manjit, *Quantum: Einstein, Bohr and the Great Debate about Reality* (UK: Icon, 2009).

Lehman, Hugh, Introduction to the Philosophy of Mathematics, APQ Library of Philosophy (Totowa: Rowan and Littlefield, 1979).

Leslie, John, Immortality Defended (MA: Blackwell Publishing, 2007).

Leslie, John, Universes (New York: Routledge 1996).

Loux, Michael, *Metaphysics: A Contemporary Introduction* (New York: Routledge, 2006).

Malin, Shimon, *Nature Loves to Hide: Quantum Physics and Reality, a Western Perspective* (New York: Oxford University Press, 2001).

Maudlin, Tim, Quantum *Non-Locality and Relativity: Metaphysical Intimations of Modern Physics* (Oxford: Wiley-Blackwell, 2011).

Moore, Walter, A Life of Erwin Schrödinger (Cambridge: Cambridge University Press, 1994).

Moreland, J.P., *Universals* (McGill-Queen's University Press, 2001).

Moser, Paul K., *The Elusive God: Reorienting Religious Epistemology* (New York: Cambridge University Press, 2008).

Murphy, Nancey, *Anglo-American Postmodernity: Philosophical Perspectives on Science, Religion, and Ethics* (Boulder: Westview Press, 1997).

Murphy, Nancey, and Ellis, George F. R., *On the Moral Nature of the Universe: Theology, Cosmology, and Ethics* (Minneapolis: Fortress Press, 1996).

Murphy, Nancey, *Beyond Liberalism and Fundamentalism: How Modern and Postmodern Philosophy Set the Theological Agenda* (Bloomsbury T and T Clark, 1996).

Murphy, Nancey, *Reasoning and Rhetoric in Religion* (Eugene OR: Wipf and Stock, 2001).

Murphy, Nancey, *Theology in the Age of Scientific Reasoning, Cornell Studies in the Philosophy of Religion* (Ithaca: Cornell University Press, 1993).

Murphy, Nancey, Russell, Robert John, Stoeger, William R., S.J., eds., Physics and Scientific Perspectives on the Problem of Natural Evil, vol I. (Vatican State: Vatican Observatory and Berkeley: Center for Theology and the Natural Sciences, 2007).

Ney, Alyssa, and Albert, David Z, ed., *The Wave Function: Essays on the Metaphysics of Quantum Mechanics* (New York: Oxford University Press, 2013).

Norris, Christopher, *Quantum Theory and the Flight from Realism: Philosophical Responses to Quantum Mechanics* (New York: Routledge, 2000).

Omnès, Ronald, *Quantum Philosophy: Understanding and Interpreting Contemporary Science*, translated by Arturo Sangalli (Princeton: Princeton University Press, 1999).

Omnès, Ronald, *Understanding Quantum Mechanics* (Princeton: Princeton University Press, 1999).

Pais, Abraham, *Niels Bohr's Times, In Physics, Philosophy, and Polity* (Oxford: Oxford University Press, 1991).

Pannenberg, Wolfhart, Peters, Ted, ed., *Toward a Theology of Nature: Essays on Science and Faith* (Louisville: John Knox Press, 1993).

Panza, Marco and Sereni, Andrea, *Plato's Problem: An Introduction to Mathematical Platonism* (New York, Palgrave Macmillan, 2013).

Peat, David F., *Einstein's Moon, Bell's Theorem and the Curious Quest for Quantum Reality* (Illinois: Contemporary Books, 1990).

Penrose, Roger, *The Emperor's New Mind: Concerning Computers, Minds, and the Laws of Physics* (Oxford: Oxford University Press, 1989).

Penrose, Roger, *Shadows of the Mind: A Search for the Missing Science of Consciousness* (New York: Random House, 2004).

Penrose, Roger, *The Road to Reality: A Complete Guide to the Laws of the Universe* (New York: Oxford, 2013).

Peters, Ted, and Bennett, Gaymon, eds., *Bridging Science and Religion* (MN: Fortress Press, 2003).

Peters, Ted, *God as Trinity: Relationality and Temporality in Divine Life* (Louisville: John Knox Press, 1993).

Peters, Ted, and Hallanger, Nathan, ed., *God's Action in Nature's World: Essays in Honor of Robert John Russell* (Aldershot: Ashgate, 2006).

Plantinga, Alvin, *Where the Conflict Really Lies: Science, Religion, and Naturalism* (Oxford: Oxford University Press, 2011).

Plantinga, Alvin, *The Nature of Necessity* (Oxford: Clarendon Press, 1974).

Plantinga, Alving, *God, Freedom, and Evil* (Grand Rapids: Eerdmans, 1974).

Plantinga, Alvin, *Does God Have a Nature?* (Wisconsin: Marquette University Press, 1980).

Plato, *Republic*, translated by Robin Waterfield (Oxford: Oxford University Press, 2008).

Plato, *Phaedo*, translated by David Gallop (Oxford: Oxford University Press, 2009).

Plato, *Phaedrus*, translated by Robin Waterfield (Oxford: Oxford University Press, 2009).

Plato, *Meno*, translated by G. M. A. Grube (Indianapolis: Hackett Publishing, 1976).

Plato, *Parmenides*, translated by Albert Whitaker (Newburyport MA: Focus Publishing, 1996).

Plato, *Theaetetus*, translated by Robin Waterfield (London: Penguin Group, 1987).

Plato, *Timaeus*, translated by Donald J. Zeyl (Indianapolis: Hackett Publishing, 2000).

Plotinus, *The Enneads*, translated by Stephen MacKenna (New York: Larson, 1992).

Plotnitsky, Akrady, *Complementarity: Anti-Epistemology after Bohr and Derrida* (Durham: Duke University Press, 1994).

Polkinghorne, John, *The Faith of a Physicist* (Princeton: Princeton University Press, 1994).

Polkinghorne, John, ed., *Meaning in Mathematics* (New York: Oxford University Press, 2011).

Polkinghorne, John, *Reason and Reality: The Relationship Between Science and Theology* (Philadelphia: Trinity Press International, 1991).

Polkinghorne, John, ed., *The Trinity and an Entangled World: Relationality in Physical Science and Theology* (Grand Rapids: WM. B. Eerdmans, 2010).

Polkinghorne, John, *Quantum Physics and Theology: An Unexpected Kinship* (Yale University Press, 2007).

Polkinghorne, John, *Quarks, Chaos and Christianity* (New York: Crossroad Publishing, 1999).

Polkinghorne, John, *Science and Providence: God's Interaction with the World* (PA: Templeton Foundation Press, 2005).

Polkinghorne, John, ed., *The Work of Love: Creation as Kenosis* (Grand Rapids: Eerdmans, 2001).

Pollard, William, *Chance and Providence God's Action in a World Governed by Scientific Law* (New York: Charles Schribner's Sons, 1958).

Popper, Karl and John C. Eccles. *The Self and its Brain: An Argument for Interactionism* (New York: Routledge, 1977).

Proclus, *A Commentary on the First Book of Euclid's Elements*, translated by G. R.

Morrow (Princeton New Jersey: Princeton University Press, 1992).

Putnam, Hilary, *Mathematics, Matter, and Method* (New York, Cambridge University Press, 1975).

Putnam, Hilary, *Philosophy of Logic* (New York: Harper and Rowe, 1971).

Quine, W. V. O., From a Logical Point of View (Cambridge: Harvard University Press, 1980).

Quine, W. V. O., *Word and Object* (Cambridge, MA: MIT Press, 1960).

Rae, Alastair, *Quantum Physics: Illusion or Reality* (Cambridge: Cambridge University Press, 1986).

Reese, Martin, *Just Six Numbers* (New York: Basic Books, 2000).

Russell, Bertrand, *The Problems of Philosophy* (Radford VA: Wilder Publications, 2008).

Russell, Robert John, *Cosmology from Alpha to Omega: The Creative Mutual Interaction of Theology and Science* (Minneapolis: Fortress Press, 2008).

Russell, Robert John, *Cosmology, Evolution, and Resurrection Hope: Theology and Science in Creative Mutual Interaction,* Proceedings of the Fifth Annual Goshen Conference on Religion and Science, ed., by Helrich, Carl S., (Ontario: Pandora Press, 2006).

Russell, Robert John, ed., *Fifty Years in Science and Religion: Ian G. Barbour and His Legacy* (Aldershot; Ashgate, 2004).

Russell, Robert John, Stoeger, William R., S.J., Coyne, George V., eds., *Physics, Philosophy and Theology: A Common Quest for Understanding* (Vatican State: Vatican Observatory, 1988).

Russell, Robert John, Murphy, Nancey, Isham, C.J., eds., *Quantum Cosmology and the Laws of Nature: Scientific Perspectives on Divine Action* (Vatican State: Vatican Observatory and Berkeley: Center for Theology and the Natural Sciences, 1993 [Revised ed. 1996]).

Russell, Robert John, Murphy, Nancey Murphy, Peacocke, Arthur, eds., *Chaos and Complexity: Scientific Perspectives on Divine Action* (Vatican City: Vatican Observatory and Berkeley: Center for Theology and the Natural Sciences, 1995).

Russell, Robert John, Stoeger, William R., S.J., Ayala, Francisco, eds., *Evolutionary and Molecular Biology: Scientific Perspectives on Divine Action* (Vatican State: Vatican Observatory and Berkeley: Center for Theology and the Natural Sciences, 1998).

Russell, Robert John, Murphy, Nancey, Meyering, Theo C., Arbib, Michael A., eds., *Neuroscience and the Person: Scientific Perspectives on Divine Action* (Vatican State: Vatican Observatory and Berkeley: Center for Theology and the Natural Sciences, 1999).

Russell, Robert John, Clayton, Philip, Wegter-Mcnelly, Kirk, Polkinghorne, John, eds., *Quantum Mechanics: Scientific Perspectives on Divine Action* (Vatican State:

Vatican Observatory and Berkeley: Center for Theology and the Natural Sciences, 2001).

Russell, Robert John, Murphy, Nancey, Stoeger, William R., S.J., eds., *Scientific Perspectives on Divine Action: Twenty Years of Challenge and Progress* (Vatican State: Vatican Observatory and Berkeley: Center for Theology and the Natural Sciences, 2007).

Russell, Robert John, *Time in Eternity: Pannenberg, Physics, and Eschatology in Creative Mutual Interaction* (Notre Dame: University of Notre Dame Press, 2012).

Sachs, Mendel, *Einstein Versus Bohr: The Continuing Controversy in Physics* (La Salle Illinois: Open Court Publishing, 1988).

Saunders, Nicholas, *Divine Action and Modern Science* (Cambridge: Cambridge University Press, 2002).

Searle, John, *Minds, Brains and Science* (Cambridge: Harvard University Press, 1984).

Segrè, Gino, *Faust in Copenhagen: A Struggle for the Soul of Physics* (New York: Viking, 2007).

Schrödinger, Erwin, *What is Life?* (Cambridge: Cambridge University Press, 1944).

Shapiro, Stewart, *Thinking about Mathematics: The Philosophy of Mathematics* (Oxford: Oxford University Press, 2000).

Spencer, John H., *The Eternal Law: Ancient Greek Philosophy, Modern Physics and Ultimate Reality* (Canada: Param Media Publishing, 2012).

Stapp, Henry, *Mind, Matter and Quantum mechanics,* 2nd ed.. (Berkeley: Springer, 2009).

Stenger, Victor, *Quantum Gods: Creation, Chaos, and the Search for Cosmic Consciousness*, (New York: Prometheus Books, 2009).

Styer, Daniel F., *The Strange World of Quantum Mechanics* (New York: Cambridge University Press, 2004).

Swinburne, Richard, *The Evolution of the Soul* (Oxford: Oxford University Press, 1997).

Swinburne, Richard, *The Existence of God* (Oxford: Oxford University Press, 2004).

Swinburne, Richard, *Mind, Brain, and Free Will* (Oxford: Oxford University Press, 2012).

Thomas, Owen C., ed., *God's Activity in the World: The Contemporary Problem* (Chico: Scholars Press, 1983).

Tracy, Thomas, ed., *The God Who Acts: Philosophical and Theological Explorations* (University Park: Pennsylvania State University Press, 1994).

van Inwagen, Peter, *Metaphysics* (Boulder Co: Westview Press, 2014).

van Inwagen, and Zimmerman, ed., *Persons: Divine and Human* (Oxford: Clarendon Press, 2007).

von Neumann, John, *Mathematical Foundations of Quantum Mechanics*, 1932., translated by Beyer, (Princeton University Press, 1996).

Ward, Keith, *Divine Action: Examining God's Role in an Open and Emergent Universe* (PA: Templeton Foundation Press, 2007).

Wigner, Eugene, *Symmetries and Reflections: Scientific Essays* (Indiana University Press, 1967).

Wright, N.T., *The Resurrection of the Son of God* (Minneapolis: Fortress Press, 2003).

Yolton, John W. Realism and Appearances: An Essay in Ontology (Cambridge: Cambridge University Press).

Zimmerman, and Loux, eds., *Oxford Handbook of Metaphysics* (Oxford: Oxford University Press, 2003).

Made in the USA
Monee, IL
03 December 2019